**Nomograms for steam generation and utilization**

# Nomograms for steam generation and utilization

V. Ganapathy

The Fairmont Press, Inc. • 700 Indian Trail • Lilburn, Georgia 30247

**Library of Congress Cataloging in Publication Data**

Ganapathy, V.
Nomograms for steam generation and utilization.

Includes index.
1. Steam engineering. 2. Steam engineering—Charts, diagrams, etc.
I. Title
TJ270.G36 1985 621.1'0121 84-45422
ISBN 0-88173-000-9

*Nomograms for Steam Generation and Utilization*
by V. Ganapathy



Published by The Fairmont Press, Inc.
700 Indian Trail
Lilburn, GA 30247

ISBN 0-88173-000-9 FP
ISBN 0-13-623257-4 PH



Second printing October 1986.

Printed in the United States of America.

Distributed by Prentice-Hall, Inc.
A division of Simon & Schuster
Englewood Cliffs, NJ 07632

Prentice-Hall International (UK) Limited, *London*
Prentice-Hall of Australia Pty. Limited, *Sydney*
Editora Prentice-Hall do Brasil, Ltda., *Rio de Janeiro*
Prentice-Hall Canada Inc., *Toronto*
Prentice-Hall Hispanoamericana, S.A., *Mexico*
Prentice-Hall of India Private Limited, *New Delhi*
Prentice-Hall of Japan, Inc., *Tokyo*
Prentice-Hall of Southeast Asia Pte. Ltd., *Singapore*

# Preface

FOR THE PAST TWELVE YEARS I HAVE BEEN authoring nomograms, charts and short articles of interest to steam power and process plant engineers, operators and managers. They were on topics related to fuels, combustion, heat transfer, boilers, heat recovery, energy conservation, life cycle costing and equipment sizing, and they appeared in widely circulated journals such as *Power Engineering, Plant Engineering, Power, Chemical Engineering, Oil and Gas Journal, Hydrocarbon Processing, Heating, Piping and Air Conditioning* and *Process Engineering* (UK). Several of these items were well received by readers all over the world, as evidenced by requests for tear sheets, additional information, etc. Many even have been asking me if I had a single source which is a collection of these articles.

Fortunately now I have been able to compile and organize over seventy of these charts, nomograms (nomographs) and short articles into a book, which I am sure would be of immense use to professional engineers, plant operators and managers, in their day-to-day work.

A question that arises is that with the plethora of personal computers whether nomograms and charts would lose their significance. As one who is also involved in developing BASIC programs I would disagree. Nomograms have certain features which make them valuable tools. These are:

1. Effects of variables can be seen at a glance. The entire regime of operation of a system is in front of you, which is psychologically a good feature. With simple moves of a straight edge one can study the effects of any variable. With a personal computer, programs have to be developed that bring out the effect of each variable. If the programmer misses something, the program will not be effective. Also, the results for a specific case alone are either printed out or displayed.

2. Large volumes of data can be developed in chart form. The nomograms on steam properties in chapter 2 is an excellent example of this. Just by moving a straight edge we can figure the enthalpy given steam pressure and temperature or the temperature given the other two variables. It is handy and convenient to use.

3. Formal education and experience in programming are essential if a program to suit one's requirements is to be developed. The readily available software packages may not really do what you want.

Computers are excellent tools when it comes to solving a complex problem and hence in my opinion both nomograms and computers will continue to be used by engineers and plant operators for years to come.

The book is divided into five chapters. Each has several selected charts. Each item is complete in itself with a brief theory, nomenclature, and example on use of the chart. An appendix has been added at the end for conversion of British units to metric if required. An index makes the cross-referencing easy.

I would like to thank the following journals that gave me permission to reproduce my articles that originally appeared in them:

*Power Engineering, Plant Engineering, Process Engineering, Chemical Engineering, Hydrocarbon Processing, Heating, Piping and Air-Conditioning, Power* and *Oil and Gas Journal.*

I would also like to thank the various readers who encouraged me to bring out a volume containing my charts and nomograms.

*V. Ganapathy*

# Contents

## Chapter 3 (continued)

## Chapter 5: Fluid Flow and Pressure Drop . . . . . . . . . . . . . . . . . . . . 137

# Chapter 1

# Fuels, Combustion, and Heat Recovery

# 1. Nomogram simplifies boiler combustion calculations

ENGINEERS ENGAGED IN BOILER DESIGN AND operation must know the weight of air required to burn a pound of fuel and the weight of flue gases produced per pound of fuel. These data are necessary before air heaters, ducts, and other auxiliary equipment can be selected, or heat transfer and energy balance calculations made.

The amount of excess air also must be determined and controlled if fuel economy is to be maximized and flue gas losses minimized.

### *Formulas*

Two methods can be used to determine the weight of air, $W_a$, required to burn 1 lb of fuel. Once $W_a$ is known, it is then used to calculate the weight of flue gases, $W_g$, produced per pound of fuel.

The first method—ultimate analysis—is a lengthy procedure, and requires a complicated chemical breakdown of the fuel:

$$W_a = 1.38 \propto \left[\frac{C}{12} + \frac{H}{4} + \frac{(S-O)}{32}\right]$$

where:

$C$ = carbon in fuel, percent
$H$ = hydrogen in fuel, percent
$S$ = sulfur in fuel, percent
$O$ = oxygen in fuel, percent
$\propto$ = excess air factor ($\propto = 1.30$ indicates 30 percent excess air)

The approximate method is a simpler procedure that also provides satisfactory results. Calculations are based on the $P$ value of the fuel —the minimum amount of air required by a fuel to release one million Btu when burned. Table I lists $P$ values for commonly used fuels.

In the approximate method,

$$W_a = (\propto P)\left(\frac{HCV}{10^6}\right)$$

where:

$HCV$ = higher calorific value of fuel (obtained from fuel supplier)

**Table 1. Minimum Air Required to Release 1 Million Btus from Fuel**

| *Fuel* | *(P, lb of air/million Btus)* |
|---|---|
| Bituminous Coal | 760 |
| Coke | 800 |
| Anthracite | 770-800 |
| Residual Oil | 745 |

The weight of flue gases produced will be the total weight of air and combustibles, or

$$W_g = W_a + 1 - \left(\frac{\text{percent ash}}{100}\right)$$

If the fuel is an oil, no ash is left; therefore,

$$W_g = W_a + 1$$

The excess air factor ($\propto$) is determined by dividing actual $CO_2$ present in the flue gases into the theoretical $CO_2$ percentage (the amount present at 0 percent excess air), or

$$\propto = \frac{\text{theoretical } CO_2\text{, percent}}{\text{actual } CO_2\text{, percent}}$$

Table 2 lists theoretical $CO_2$ values for several fuels.

The nomograph can be used to quickly determine $\propto$, $W_a$, and $W_g$.

**Table 2. Theoretical $CO_2$, Values for Industrial Fuels**

| *Fuel* | *Theoretical $CO_2$, percent* |
|---|---|
| Bituminous Coal | 18.6 |
| Coke | 19.7 |
| Anthracite | 19.5 |
| Residual Oil | 16.4 |
| Distillate Oil | 15.5-16 |

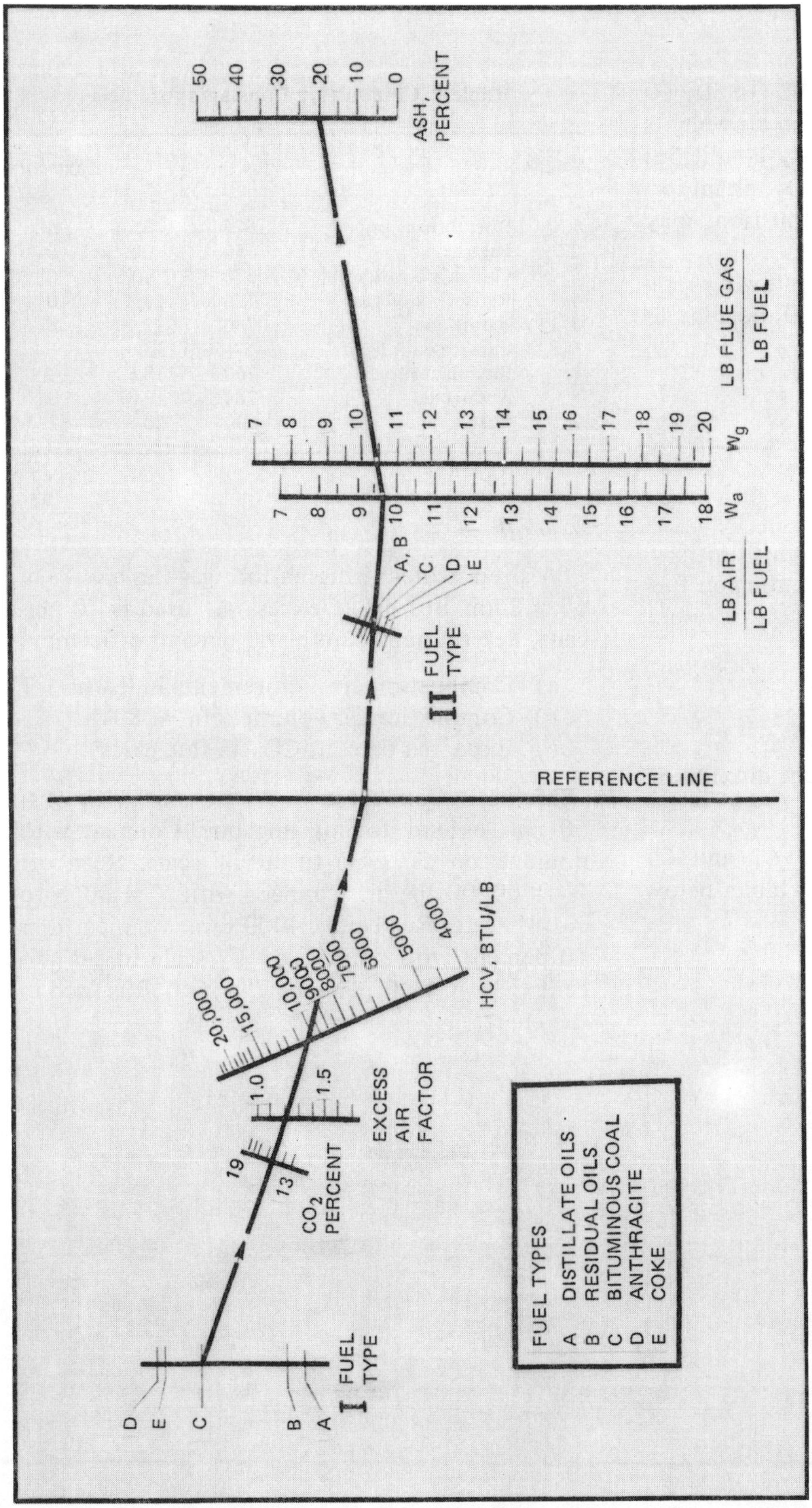


### *Example*

What are $\propto$, $W_a$, and $W_g$ when a boiler is burning bituminous coal, *HCV* is 10,500 Btu/lb, ash is 20 percent, and actual $CO_2$ is 15.5 percent?

### *Solution*

On the nomograph, connect *C* (bituminous coal) on fuel scale I with 15.5 percent on $CO_2$ scale, extend line to the excess air scale, and read $\propto$ as 1.2, or 20 percent.

Connect this point with 10,500 Btu/lb on scale *HCV* and extend line to reference line. Connect intersection point on reference line with *C* on fuel scale II, extend line to scale $W_a$, and read the weight of the air required to burn 1 lb of fuel as 9.7 lb.

Then, connect 9.7 on scale $W_a$ to 20 on ash scale. Where this line intersects scale $W_g$, read 10.5 lb of flue gas produced per pound of coal burned.

The procedure is the same if the fuel is an oil. The line from scale $W_a$ would be connected with 0 on the ash scale to determine $W_g$. Or, the same answer can be obtained by adding 1 to $W_a$.

# 2. Chart relates furnace duty with air flow and %$CO_2$ in flue gas

THE FOLLOWING CHART MAY BE USED TO obtain good estimates of combustion air weights and volume at various furnace duty conditions and excess air. Also, the percent $CO_2$, an important variable for monitoring combustion, may be obtained from the same chart.

The charts are based on the following relations. A furnace efficiency of 80 percent is assumed.

$$W = 1.25\, Q\, C_1 \alpha \quad (1)$$

$$V = W(460 + T)/2361 \quad (2)$$

$$\text{percent } CO_2 = C_2/\alpha \quad (3)$$

where:

$Q$ = furnace duty, million Btu/hr

$C_1, C_2$ are constants depending on fuel type

$\alpha$ = excess air factor

$W$ = air quantity, lbs/hr

$V$ = air volume, cfm

$T$ = temperature of air, °F

percent $CO_2$ = volume of carbon dioxide in dry flue gas

Fuel type is indicated on the $C_1$ and $C_2$ scales directly by symbols. The table below gives the values of $C_1$, $C_2$ used.

**Table 1. Combustion Constants for Fuels**

| *No.* | *Fuel* | $C_1$ | $C_2$ | *Symbol* |
|---|---|---|---|---|
| 1. | Blast furnace gas | 575 | 25 | A |
| 2. | Bagasse | 650 | 20 | B |
| 3. | Carbon monoxide gas | 670 | 10 | C |
| 4. | Refinery & oil gas | 720 | 13.3 | D |
| 5. | Natural gas | 730 | 12.0 | E |
| 6. | Furnace oil & lignite | 745-750 | 15.0 | F |
| 7. | Bituminous coals | 760 | 18.6 | G |
| 8. | Anthracite | 780 | 19.5 | H |
| 9. | Coke | 800 | 20.5 | I |

***Example***

A fired heater firing natural gas has a duty of 60 million Btu/hr. If excess air used is 10 percent, determine, assuming 80 percent efficiency:

a) Combustion air requirements in lbs/hr
b) Combustion air volume, cfm at 80°F
c) Expected percent $CO_2$ in flue gases

Method: Connect $\alpha$ = 10 percent with $Q$ = 60 and extend to cut line one. Connect with point E on $C_1$ scale to cut $W$ scale. Read off $W$ = 60200 lbs/hr. Connect with $T$ = 80°F to cut $V$ scale. Read off 14000 cfm. Connect $\alpha$ = 10 percent with point E on $C_2$ scale to get percent $CO_2$ = 11.0 percent. $T$ = 540°R is used in the chart.

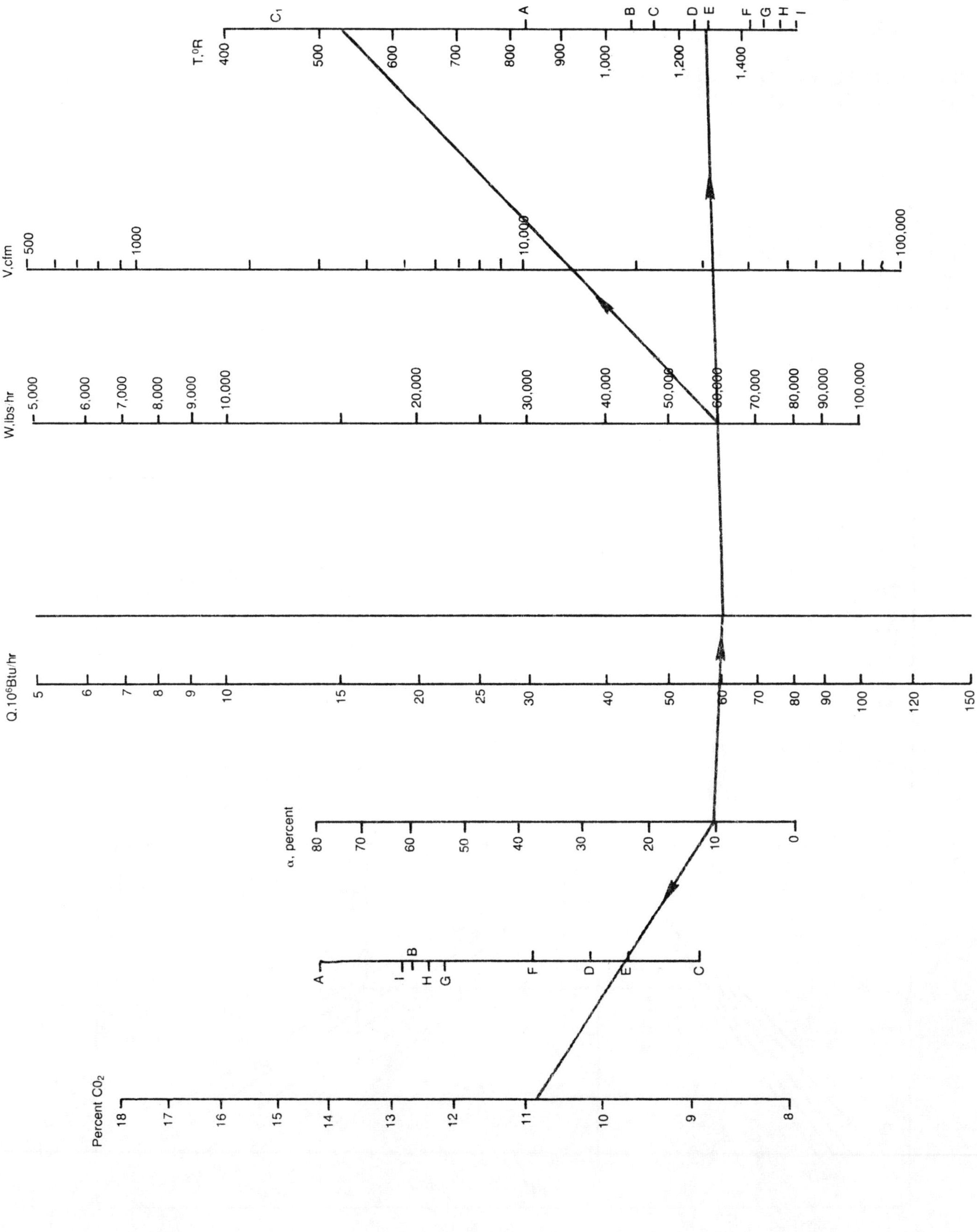


Figure 1. Furnace Parameter Chart

# 3. Nomogram estimates combustion air requirements for various fuels

THE AIR-FUEL RATIO VARIES CONSIDERABLY for the different fuels used in boilers, and this nomogram is useful in obtaining a quick estimate of air requirements. Although the air requirements for natural gas and oil are not identical, they are reasonably close, and the single scale used in the nomogram provides a good approximation. Similarly, air required by bituminous coals varies with volatile content, but the single scale again provides a good approximation.

### *Example*

A 250-MW steam generator in a power plant having a plant heat rate of 8500 Btu/kwh is fired with natural gas at 25% excess air. Air temperature is 500F. Find the weight and volume of air required.

### *Solution*

As indicated, go vertically from unit rating to heat rate, horizontally to type of fuel, vertically down to percent excess air, horizontally over to air temperature, and vertically up to volume scale. Read off 2.3 million lb/hr and 900,000 cfm at 500F.

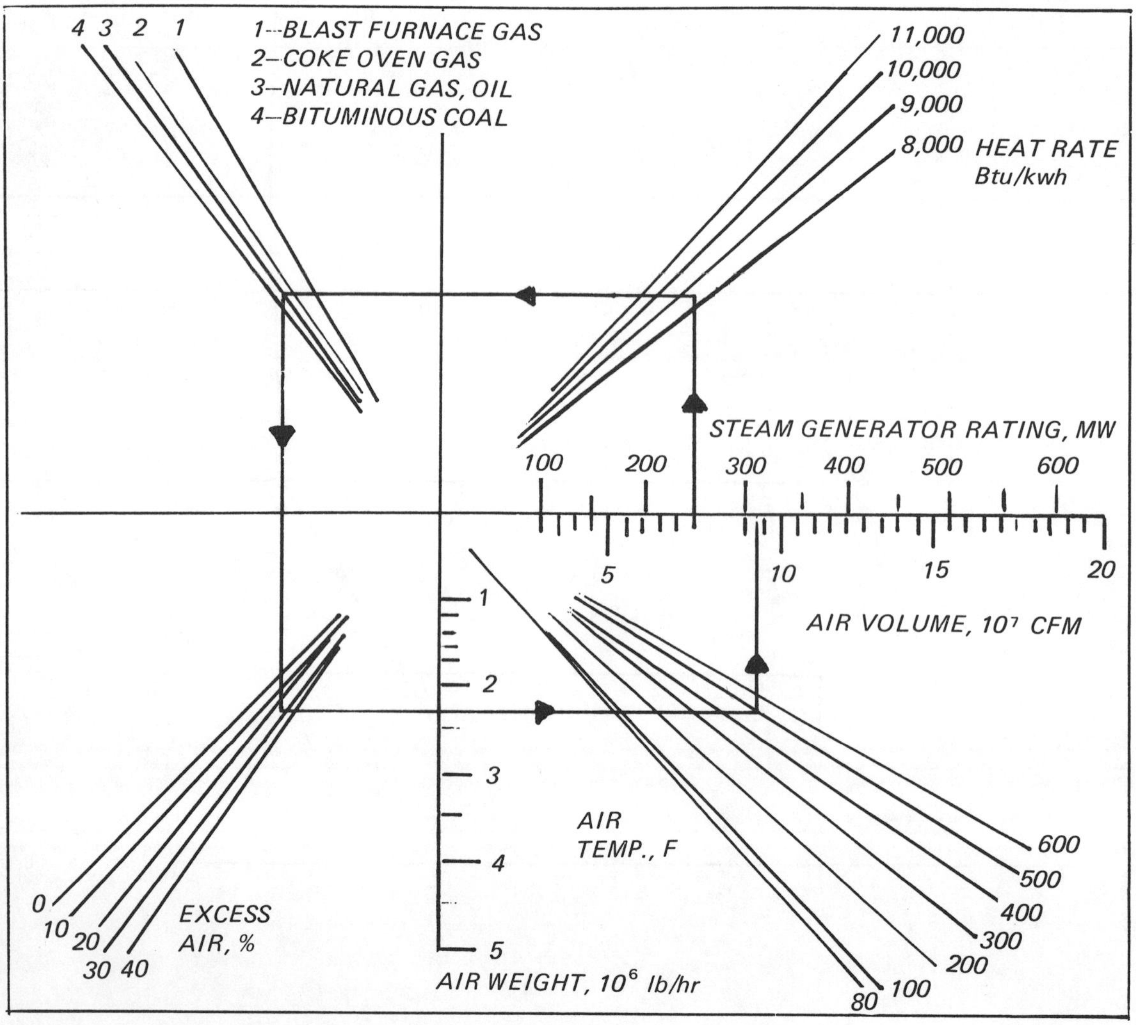

# 4. Chart aids quick comparisons of energy costs

ENGINEERS ARE OFTEN CONCERNED WITH THE comparative costs of various fuels and electricity. The question often is which fuel or electricity is most economical.

The chart helps answer the question quickly. It gives the cost per million Btu of solid, liquid, and gaseous fuels. It also gives the energy cost in \$/MMBtu and cents/kw-hr so a comparison can be made between fuel and electricity. Various fuels and their range of heating values are indicated. Table 1 also gives the heating value of various fuels.

### *Example 1*

Coal with a higher heating value (HHV) of 9,000 Btu/lb costs \$18/ton. What is the cost in \$/MMBtu?

**Table 1. Higher Heating Values of Fuels**

| *SOLID FUELS* | *HHV, Btu/lb* |
|---|---|
| Low volatile bituminous coal and high volatile anthracite coal | 11,000-14,000 |
| Subbituminous coal | 9,000-11,000 |
| Lignite | 7,000- 8,000 |
| Coke and coke breeze | 12,000-13,000 |
| Wood | 8,000- 9,500 |
| Bagasse (dry) | 7,000- 9,000 |
| *LIQUID FUELS* | *HHV, Btu/gal.* |
| Fuel oil | |
| No. 1 | 137,400 |
| No. 2 | 139,600 |
| No. 3 | 141,800 |
| No. 4 | 145,100 |
| No. 5 | 148,800 |
| No. 6 | 152,400 |
| Gasoline (60° API) | 150,000 |
| Kerosine (45° API) | 159,000 |
| Gas oil (30° API) | 16,907 |
| Fuel oil (15° API) | 178,710 |
| *GASEOUS FUELS* | *HHV, Btu/cu ft* |
| Coke oven gas | 500- 700 |
| Blast furnace gas | 90- 120 |
| Natural gas | 900-1,300 |
| Refinery off gas | 1,100-2,000 |

### *Solution*

In the chart, connect \$18/ton on C scale with 9,000 Btu/lb on HHV scale, extend to cut the F scale, and read \$1/MMBtu.

### *Example 2*

An engineer has to choose between No. 6 fuel oil which costs 25¢/gal. and coal of a HHV of 9,000 Btu/lb that costs \$25/ton. Which should he use?

### *Solution*

Connect \$0.25 on the C scale with point C (No. 6 fuel oil) on the HHV scale and extend to cut the F scale at \$1.7/MMBtu. Next, connect 1.7 on the F scale with 9,000 on the HHV scale, extend to cut the C scale, and read \$30/ton for equivalent coal cost.

Since the coal costs \$25/ton, it is more economical to use coal. However, other costs associated with coal combustion should be worked out before a decision is made.

### *Example 3*

Natural gas with a higher heating value of 1,050 Btu/cu ft costs 75¢/1,000 cu ft. What is the cost in \$/MMBtu?

### *Solution*

Connect \$0.75 on the C scale with G on the HHV scale, and extend to cut the F scale at \$0.72/MMBtu.

Note that the letter symbols on the HHV scale, identified by the legend, for various liquid and gaseous fuels may be used if the exact heating value is not known. If the heating value is known, use the following procedure to get more accurate results.

For gaseous fuels, multiply the heating value by 10 and use the product on the HHV scale. For example, a natural gas has a heating value of 1,050 Btu/cu ft. Thus 10 × 1,050 = 10,500 gives the value to use on the HHV scale.

For liquid fuels, divide the heating value in

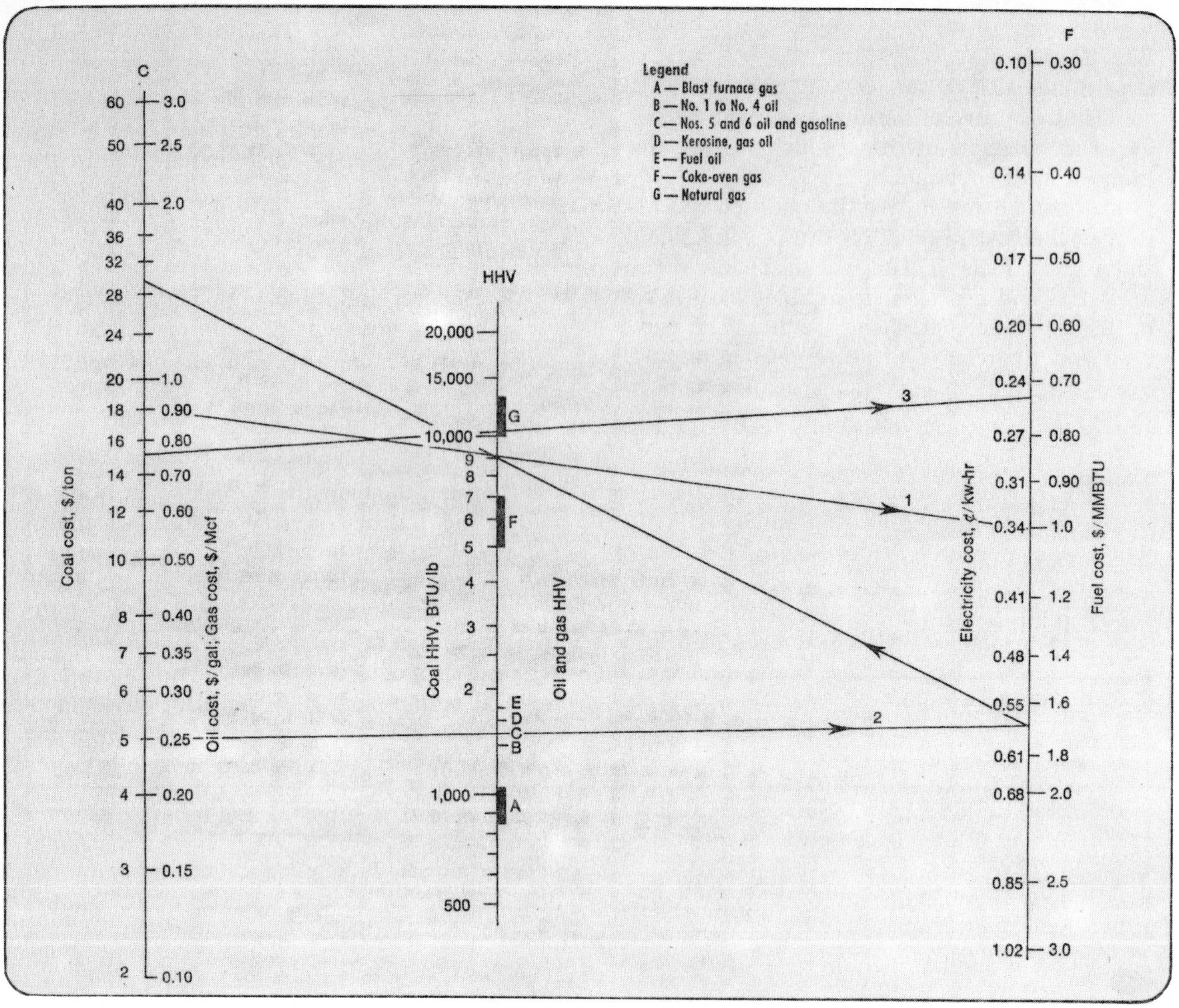


Btu/gal. by 100 and use it on the HHV scale. For example, No. 6 fuel oil has a heating value of 152,000 Btu/gal. Thus, 152,000/100 = 1,520 is the value to use on the HHV scale.

Thus by using the foregoing procedures, any solid, liquid, or gaseous fuel may be considered.

Conversion of energy costs from $/MMBtu to ¢/kw-hr can be made by using the F scale. For example, $1.8/MMBtu equals 0.61¢/kw-hr. Thus one can check to see if electricity is more economical than fuel.

### *Example 4*

An engineer has the option of using natural gas with a HHV = 1,050 Btu/cu ft at $0.75/1,000 cu ft or electricity at 1.2¢/kw-hr. Compare the two to see which is most economical.

### *Solution*

From Example 3, fuel costs $0.72/MMBtu. This equals 0.25¢/kw-hr. But electricity costs 1.2¢/kw-hr. Thus, natural gas is the economical choice. Of course using natural gas involves such capital expenditures as burners, etc. Still, the chart quickly gives an idea of the comparative energy costs.

The following equations were used in preparation of the chart:

$$F_c = 500\, C_c / \mathrm{HHV}_c \tag{1}$$

$$F_g = 1{,}000\, C_g / \mathrm{HHV}_g \tag{2}$$

$$F_o = 10^6\, C_o / \mathrm{HHV}_o \tag{3}$$

$$E = 0.3413\, F \tag{4}$$

where:

$C_c$, $C_g$, and $C_o$ are the costs of solid, gaseous, and liquid fuels in \$/ton, \$/1,000 cu ft, and \$/gal., respectively. $F_c$, $F_g$, and $F_o$ are the costs of coal, gaseous fuels, and liquid fuels in \$/MMBtu.

$\mathrm{HHV}_c$, $\mathrm{HHV}_g$, and $\mathrm{HHV}_o$ are the higher heating values of solid, gaseous, and liquid fuels in Btu/lb, Btu/cu ft, and Btu/gal., respectively.

$E$ is the equivalent \$/MMBtu in ¢/kw-hr.

# 5. Chart estimates heat loss in flue gases

THIS NOMOGRAPH PROVIDES a quick method for estimating heat losses in flue gases discharged from boilers and process furnaces fired by natural gas, fuel oil, or bituminous coal. Oxygen content and temperature of the flue gases, and ambient air temperature, must be measured before applying the illustration. The heat loss percentage indicated on the nomograph is based on the lower calorific heating value of the fuel.

### *Problem*

A boiler fired by fuel oil is discharging 500F flue gases containing 2 percent oxygen into 100F ambient air. How much heat is escaping with the flue gases?

### *Solution*

Connect 500F on Scale G with 100F on Scale A; then extend a vertical line down to the fuel oil curve (no. 2) on Scale F. Connect the marked point on Scale F with 2 percent on Scale P; then extend a vertical line up to Scale L and read the heat loss as 8.3 percent.

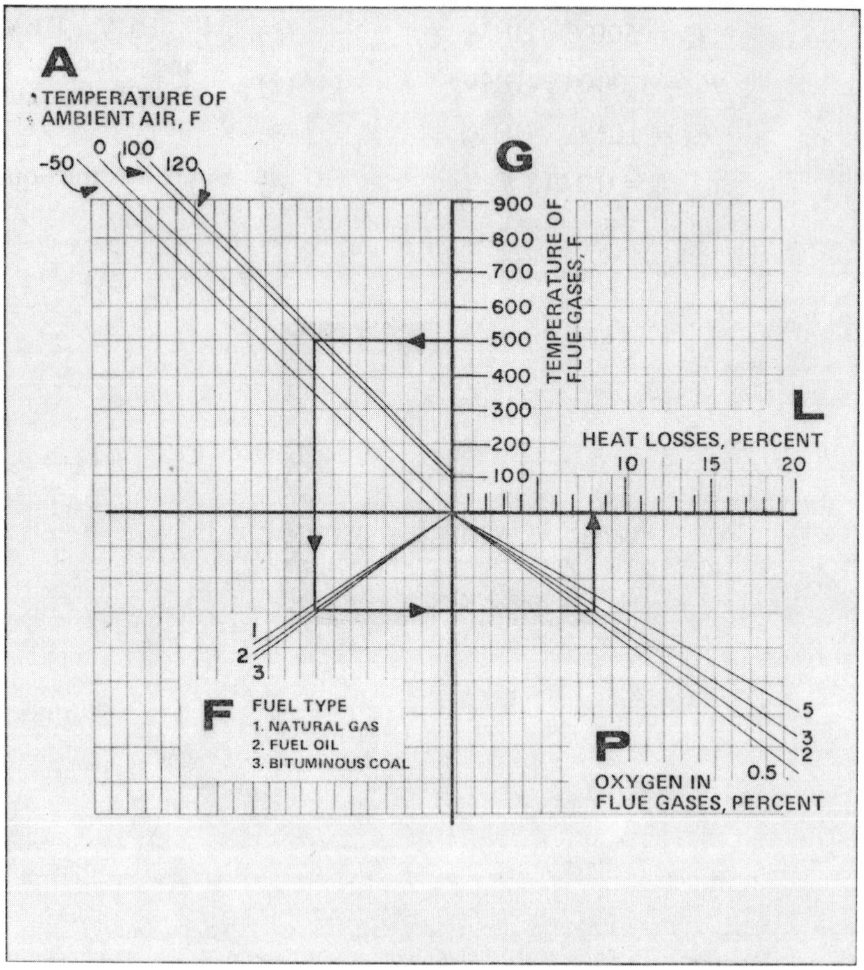

# 6. Chart speeds up computation of flue-gas loss

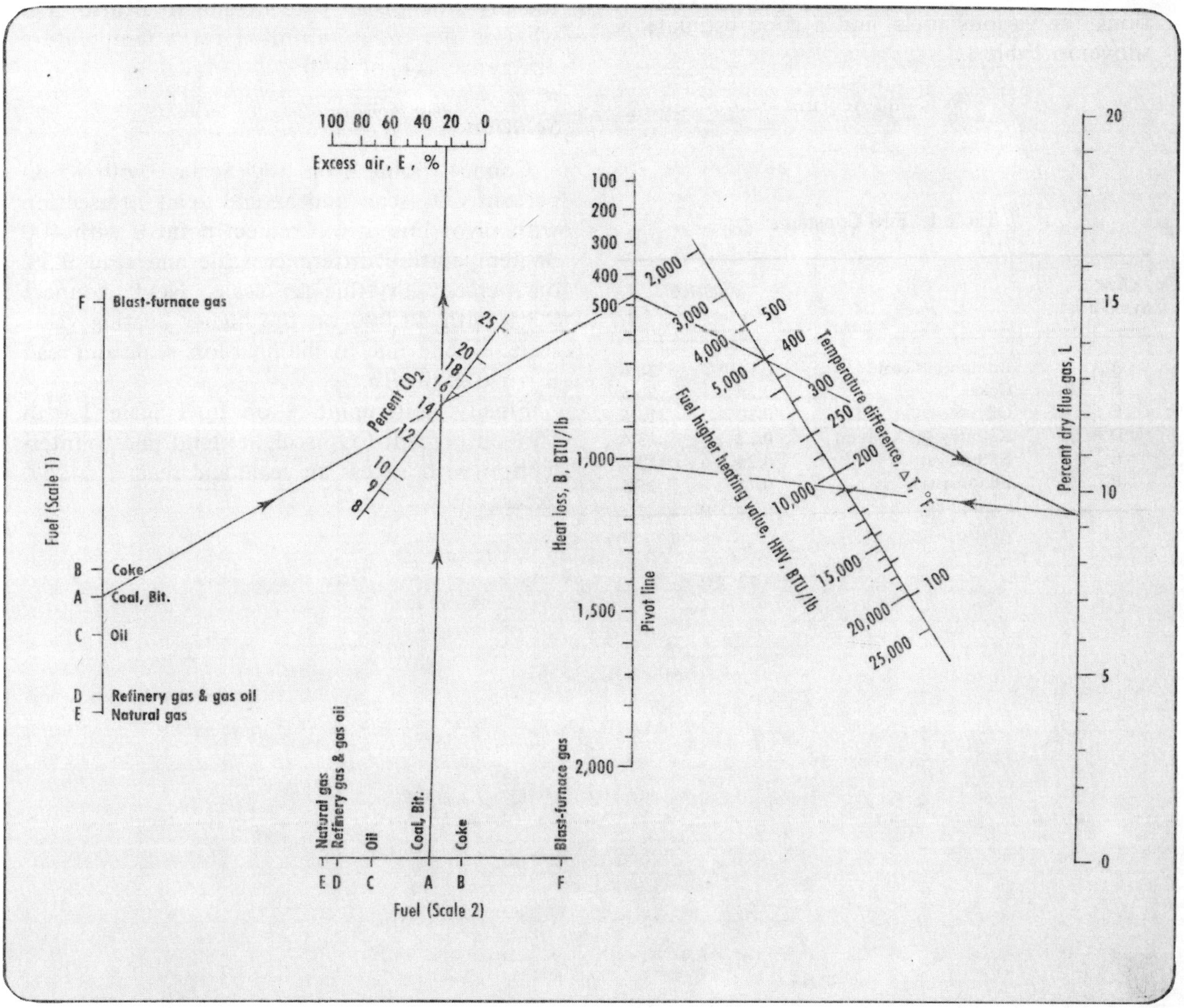


THE ALIGNMENT CHART OR NOMOGRAM presented here provides a simple way to compute the percent dry flue-gas loss with various fuels. If the percent $CO_2$ and the stack temperature are known, the chart provides accurate enough results for engineering estimates.

The formulas on which the chart is based are:

Percent dry flue gas,

$$L = C_1 \Delta T / \%CO_2 \quad \text{(1)}$$

Dry flue-gas loss, Btu/lb,

$$B = (L/100)HHV \quad \text{(2)}$$

Excess-air factor,

$$E = C_2 / \%CO_2 \quad \text{(3)}$$

In Equation 1, $\Delta T$ equals the difference in the gas temperature leaving the stack and the air temperature entering.

Thus, if percent $CO_2$ is known, excess air also can be calculated.

Too, the dry gas loss, Btu/lb, can be obtained if the heating value (HHV) of the fuel is known.

Constants $C_1$ and $C_2$ for Equations 1 and 3 have been developed from combustion calculations for various fuels, and a good estimate is shown in Table 1.

**Table 1. Fuel Constants**

| *Chart Symbol* | *Fuel* | *Constants* $C_1$ | $C_2$ |
|---|---|---|---|
| A | Bituminous coal | 0.35 | 18.6 |
| B | Coke | 0.39 | 20.5 |
| C | Oil | 0.30 | 15.5 |
| D | Refinery gas & gas oil | 0.22 | 13.4 |
| E | Natural gas | 0.20 | 12.5 |
| F | Blast-furnace gas | 0.75 | 25.5 |

### *Example*

Coal with a HHV of 11,000 Btu/lb is fired in a boiler. The percent $CO_2$ is read as 15. What is the dry flue gas in percent and in Btu/lb, and what is the excess air used for a temperature difference, $\Delta T$, of 400F?

### *Solution*

Connect point A on Fuel Scale 1 with 15 on percent $CO_2$ scale and extend to an intersection with pivot line at 0. Connect point 0 with 400 on temperature difference scale and read 9.3% on percent dry-flue-gas scale. Next, connect 9.3% with 11,000 on the higher heating value scale, extend line to the heat-loss scale and read B = 1,020 Btu/lb.

Finally, join point A on Fuel Scale 2 with 15% on percent $CO_2$ scale, extend line to intersection with excess air scale and read E = 25% excess air.

# 7. Nomogram estimates efficiencies of boilers and fired heaters

THIS NOMOGRAM PERMITS QUICK ESTIMATES of the efficiency of boilers and fired heaters based on the lower heating value of the fuel. It is valid for coals, oil, and natural gas.

The weight of air required to burn one million Btu input of fuel may be written as:

$$A = C(1 + E/100)$$

where:

$E$ = excess air, %
$C$ = constant, varies from 730 to 760 for fuels on a higher heating value basis (Ref. 1)

Flue gas produced per pound of fuel may then be written as:

$$w = C(1 + E/100)(H/10^6) + 1$$

where:

$H$ = higher heating value, Btu/lb

Wet flue gas loss ($X$, in %) on a lower heating value basis may be written as:

$$X = 0.24\, w(t_g - t_a)100/L$$

where:

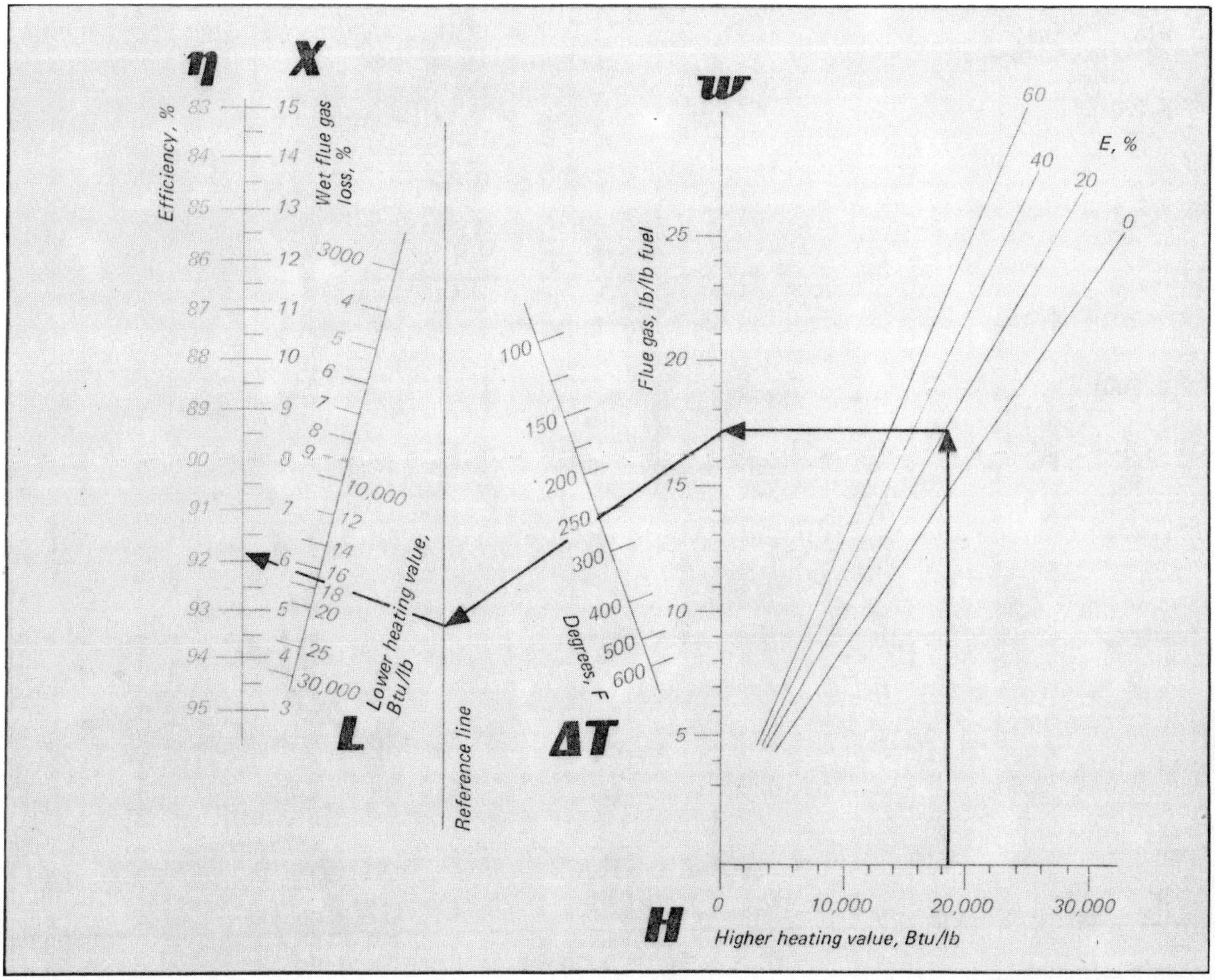

$t_a$ = ambient air temperature, F
$t_g$ = exit gas temperature, F
$L$ = lower heating value, Btu/lb

The total loss is the sum of flue gas loss, radiation loss, and unaccounted losses; so, efficiency ($\eta$ in %) is: $\eta = 100 - X - (\text{rad.} + \text{unacc.})$.

To use the nomogram it is necessary to know higher and lower heating values of the fuel, amount of excess air, and the exit gas and ambient air temperatures.

### *Example*

Fuel oil having higher and lower heating values of 18,000 and 17,000 Btu/lb is fired with 20% excess air. If $t_g$ = 350F and $t_a$ = 100F, find flue gas produced *(w)*, wet flue gas loss *(X)*, and efficiency ($\eta$).

### *Solution*

Go vertically from $H$ = 18,000 to cut the $E$ = 20 line. Go horizontally from this point to cut the $w$-scale, and read $w$ = 17.2 lb flue gas/lb fuel. Align this point with $t_g - t_a$ = 250 and extend to reference line. Align this point with $L$ = 17,000, and extend to cut $X$-scale. Read $X$ = 6.2%. The efficiency scale is based on the assumption of 2% additional losses; so the efficiency on lower heating value basis is 91.8%. This can be converted to higher heating value basis by multiplying by the ratio of heating values: Eff. on HHV basis = 91.8 (17,000)/18,000 = 86.7%.

### *Reference*

V. Ganapathy, *Applied Heat Transfer,* Pennwell Books, Tulsa, 1982.

# 8. Nomogram estimates combustion-gas dewpoint

A STARTING POINT IN THE DESIGN OR SELECTION of low-temperature heat-recovery surfaces such as economizers or air heaters is a knowledge of the dewpoint of flue gases, above which the heat-recovery surfaces should be maintained to avoid corrosion.

The alignment chart (below) gives a good estimate of dewpoint temperature ($T_{DP}$) based on the equation: (Ref. 1)

$$1{,}000/T_{DP} = 1.7842 + 0.0269 \log P_{H_2O} - 0.129 \log P_{SO_3} + 0.329 \log P_{H_2O} \times \log P_{SO_3}$$

$P_{H_2O}$ and $P_{SO_3}$ are partial pressures of water vapor and sulfurtrioxide, in atmospheres. $T_{DP}$ is dewpoint in °K.

Estimates of $SO_3$ in flue gases for oil and coal are given in Table 1 for various excess-air and

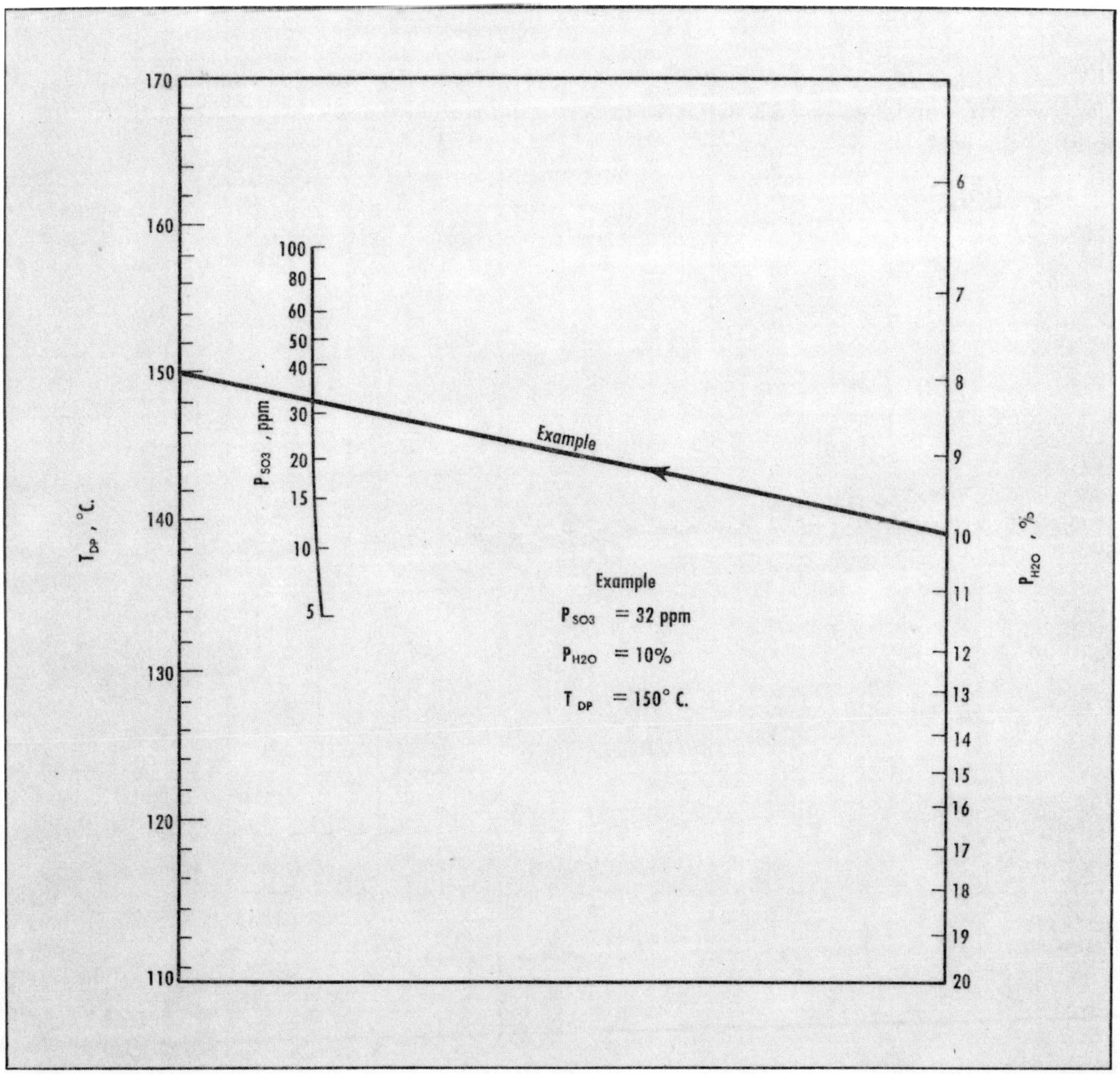

**Table 1. $SO_3$ in Flue Gas**

| Fuel | Excess air, % | Sulfur, % 0.5 | 1.0 | 2.0 | 3.0 | 4.0 | 5.0 |
|---|---|---|---|---|---|---|---|
| | | $SO_3$, ppm | | | | | |
| Oil | 5 | 2 | 3 | 3 | 4 | 5 | 6 |
| | 11 | 6 | 7 | 8 | 10 | 12 | 14 |
| Coal | 25 | 3-7 | 7-14 | 14-28 | 20-40 | 27-54 | 33-66 |

sulfur contents. As an example, oil fired with 0.5% S at 5% excess air gives 2 ppm of $SO_3$. Alignment chart gives $T_{DP}$ in °C when $P_{SO3}$ is in ppm and $P_{H20}$ is in percentage by volume.

*Example*

If $P_{H20}$ = 10% and $P_{SO3}$ = 32, what is the dewpoint of the gas?

*Solution*

Connect $P_{H20}$ = 10% with $P_{SO3}$ = 32 ppm, extend to the $T_{DP}$ scale, read 150°C as gas dewpoint.

*Reference*

1. Pierce, Robert, "Estimating Acid Dew Point in Stack Gases," *Chemical Engineering,* April 1977.

# 9. Nomogram estimates particulate collection efficiency for coal-fired boiler emissions

HERE IS A QUICK, GRAPHICAL METHOD FOR determining pounds of particulate emissions discharged from coal-fired boilers and the efficiency required of an attached dust collection system for the boiler to meet allowable pollution control levels. Equations represented in the nomograph are:

$$\text{Actual emission rate} = \frac{8500 \times \text{percent ash in coal}}{\text{higher calorific value}}$$

$$\text{Collection efficiency} = \left(1 - \frac{\text{allowable rate}}{\text{actual rate}}\right) \times 100$$

***Problem***

Coal with a higher calorific value (HCV) of 11,500 Btu/lb and 12-percent ash is fired in a boiler. What is the particulate emission rate in lb/million Btu? If local EPA regulations allow 0.2 lb/million Btu, what is the required efficiency of the dust collection system? (It is assumed that 85 percent of the ash in the coal leaves the boiler as flyash.)

***Solution***

Draw line from 12 on scale A to 11,500 on scale B. At point where line intersects scale C, read particulate emission rate as 8.9 lb/million Btu.

Connect this intersection point with 0.2 on scale D, extend line to scale E, and read required efficiency for dust collection system as 97.8 percent.

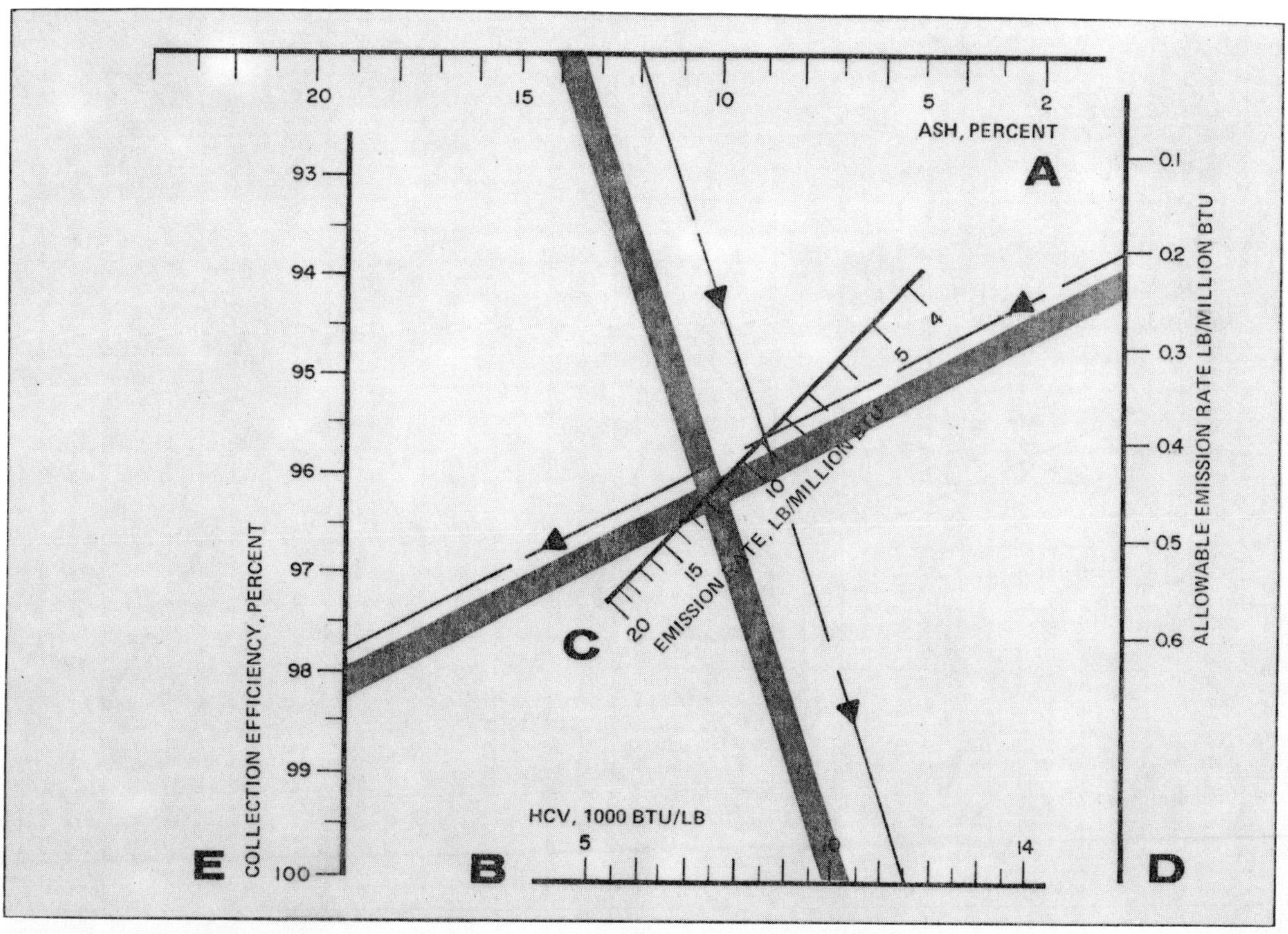

# 10. Nomogram estimates particulate emission rates

PARTICULATE-EMISSION DATA ARE NEEDED TO size dust collectors for coal-fired boilers. The amount of particulate matter is typically expressed in grains/ft³, and is a function of many variables such as fluegas temperature, caloric value and percentage ash in the coal, and excess air. This chart speeds up the calculation.

Dust concentration in the fluegas ($C$, grains/ft³) is described by the following equation for boilers fired with pulverized coal:

$$C = \frac{240{,}000\,(\%\text{ash}/100)}{T\,[7.6\times10^{-6}\,(HHV)\,(100+\alpha)+1-(\%\text{ash}/100)]}$$

where:

$T$ = fluegas temperature, °R(°F+460)
$\alpha$ = percent excess air
$HHV$ = higher heating value of coal, Btu/lb

This is the equation used in the chart. For conversion to other concentration units, note that 1 grain/ft³ = 2.2884 g/m³ and that 1 lb = 7,000 grains. Also be aware that concentration is based on actual (as opposed to standard) fluegas volume.

***Example***

What is the concentration of particulate matter in the fluegas from a boiler firing coal with 11,000 Btu/lb heating value and 11% ash? Excess air is 25% and fluegas temperature is 390°F (850°R).

***Solution***

On the chart, connect excess air = 25% with $HHV$ = 11,000 and extend to cut line 1. Connect this point with %ash = 11 and extend to cut line 2. Connect this point with $T$ = 850°R and extend to read off 2.75 grains/ft³ as the dust concentration in the fluegas.

What if the coal had a heating value of 10,000 Btu/lb and was 12% ash? Then the chart would show that dust in the fluegas is 3.25 grains/ft³.

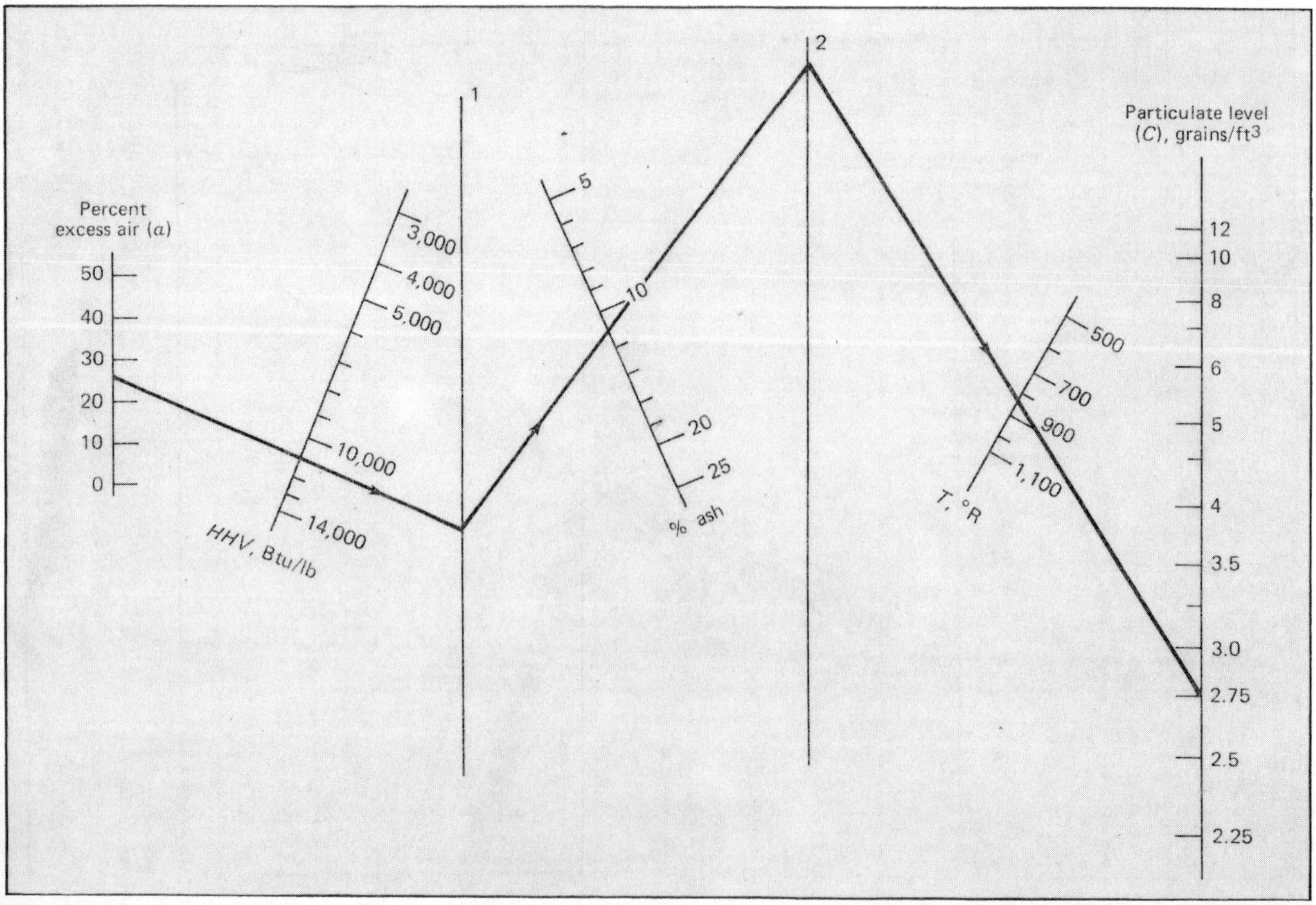

# 11. Nomogram estimates $SO_2$ emission rates

THE ACCOMPANYING NOMOGRAM OR ALIGNMENT chart gives stack sulfur dioxide ($SO_2$) emissions (A) in lb/MMBtu fired based on the percent sulfur in the fuel—either coal or oil—and on the higher heating value (HHV). It also gives the efficiency of the $SO_2$ removal system required (P) to limit $SO_2$ emissions to a particular value (C).

The chart is based on the equations,

$$A = 2 \times 10^4\ S/HHV$$
$$P = 100(1 - C/A)$$

*Example*

When coal with 3% sulfur and 9,500 Btu/lb HHV is fired, what is the $SO_2$ emissions and the efficiency of the removal system required to limit emissions to 1.0 lb/MMBtu?

*Solution*

Connect 3 on the sulfur-content scale (S) with 9,500 on the HHV scale, extend to the stack emissions scale (A), and read $SO_2$ = 6.3 lb/MMBtu. Next, connect the 6.3 on the (A) scale with 1 on the $SO_2$ emissions limit (C) scale and read 84% on the $SO_2$ removal system efficiency (P) scale.

Note that the sulfur content (S), $SO_2$ emissions (A), and $SO_2$ emissions limit (C) scales are linear and may be extrapolated.

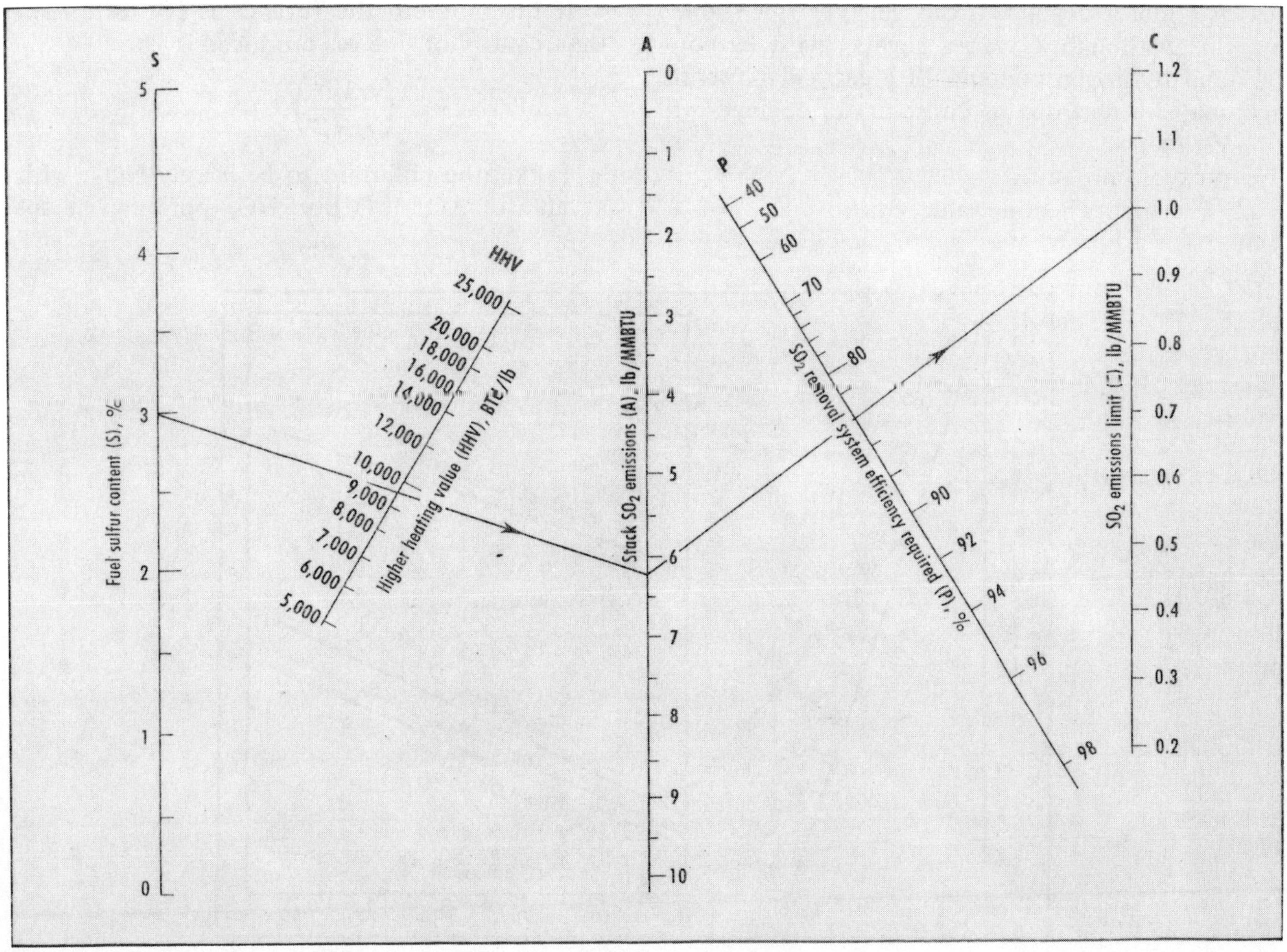

# 12. Chart aids easy conversion of NOx units

AMONG THE MORE COMMON POLLUTANTS required to be controlled are the various oxides of nitrogen, widely known as $NO_x$.

Pollution limits are normally expressed in either volume or mass units and it is often desirable for engineers to convert from one unit to another with ease. Units used are ppm and lb/MBtu and a relation between these can be derived.

If $W_g$ is the quantity of flue gas produced per lb of fuel fired, then it may be shown from combustion calculations[1,2] that:

$$W_g = \left(\frac{C\alpha HHV}{10^6}\right) + \left(1 - \frac{\%\text{ash}}{100}\right) \qquad (1)$$

where

$C$ is a factor dependent on fuel type (see Table 1). Although $C$ varies slightly with carbon and hydrogen content for a particular fuel it may be regarded as constant for engineering estimates.

$\alpha$ = excess air factor

$HHV$ = higher heating value, Btu/lb

**Table 1. Typical Fuel Data**

| *No.* | *Type of fuel* | *HHV, Btu/lb* | *C, lb/ MBtu* | *Mol wt of flue gas* |
|---|---|---|---|---|
| 1. | Fuel oil | 19,500 | 745 | 39 |
| 2. | Natural gas | 20,000 | 730 | 30 |
| 3. | Coal (8% ash) | 12,000 | 760 | 30 |
| 4. | Blast Furnace gas | 1,200 | 575 | 31.5 |
| 5. | Refinery and oil gas | 22,000 | 720 | 30 |
| 6. | Coke oven gas | 20,000 | 670 | 28 |

For oil and gaseous fuels, % ash is negligible and hence:

$$W_g = \left(\frac{C\alpha HHV}{10^6}\right) + 1 \qquad (2)$$

If the input to the furnace is $10^6$ Btu, then the quantity of flue gas produced is

$$\left(\frac{10^6}{HHV}\right) \times W_g$$

Taking the pollutant to be largely $NO_2$, with a value of $NO_x$ lb/MBtu, $NO_x$ ppm can be ap-

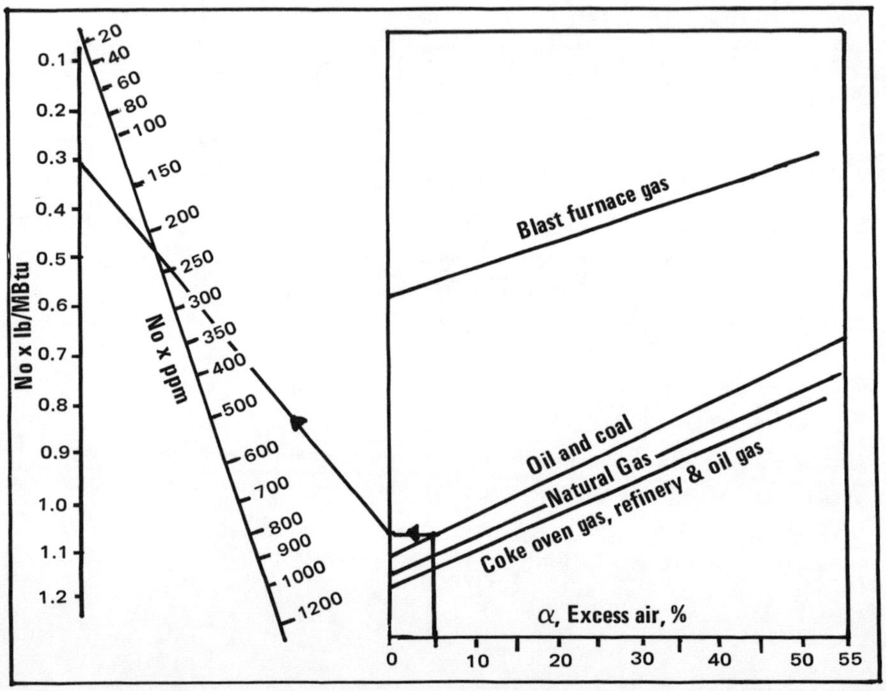

proximated by equation (3) using partial pressure laws:

$$NO_x \text{ lb/MBtu} = NO_x \text{ ppm}\left(\frac{46}{M_g}\right)\left(\frac{W_g}{HHV}\right) \quad (3)$$

where 46 is the molecular weight of $NO_2$ and $M_g$ is the molecular weight of the flue gas.

### *Estimations*

$M_g$ varies with the type of fuel fired and the excess air used. However, for estimation purposes, the values in Table 1 have been used. Equation 3 may be simplified and is shown as the nomograph.

The following calculation shows the effect of *HHV* on $NO_x$ is insignificant for a particular fuel as it occurs in the numerator, in $W_g$, and in the denominator. Say two fuels have *HHVs* of 18,000 and 20,000 Btu/lb respectively. The factor $W_g/HHV$, when calculated for 10% excess air using equation 1, is $8.75 \times 10^{-4}$ and $8.69 \times 10^{-4}$ respectively, a variation of 0.6 to 0.7% for a change in *HHV* of 11%. Hence, differences in *HHV* can be neglected.

### *Example*

When fuel oil is fired with 5% excess air and the pollution limit is 0.3 lb/MBtu $NO_x$, the volume of $NO_x$ in the wet flue gas is 230 ppm.

It is common to express the volume of $NO_x$ in ppm dry gas. The relation between the volume of $NO_x$ in the dry and wet gases is simple, only involving a ratio. If the volume of water vapour in the flue gas is $V\%$.

$$NO_x \text{ dry} = NO_x \text{ wet} \times \left(\frac{100}{100-V}\right) \quad (4)$$

If moisture is given as a percentage by weight, $Y\%$, then the following relationship can be derived, assuming a molecular weight of 31 for dry gases:

$$NO_x \text{ dry} = NO_x \text{ wet} \times \left(\frac{1+0.72Y}{1-Y}\right) \quad (5)$$

### *References*

1. Ganapathy, V. Simplifying boiler combustion calculations, *Plant Engineering,* 9 Jan 1975, pp 56–57.
2. Ganapathy, V. Basic data for steam generators at a glance. *Chemical Engineering,* 6 June 1977, pp 197-198.

# 13. Nomogram estimates melting point of ash

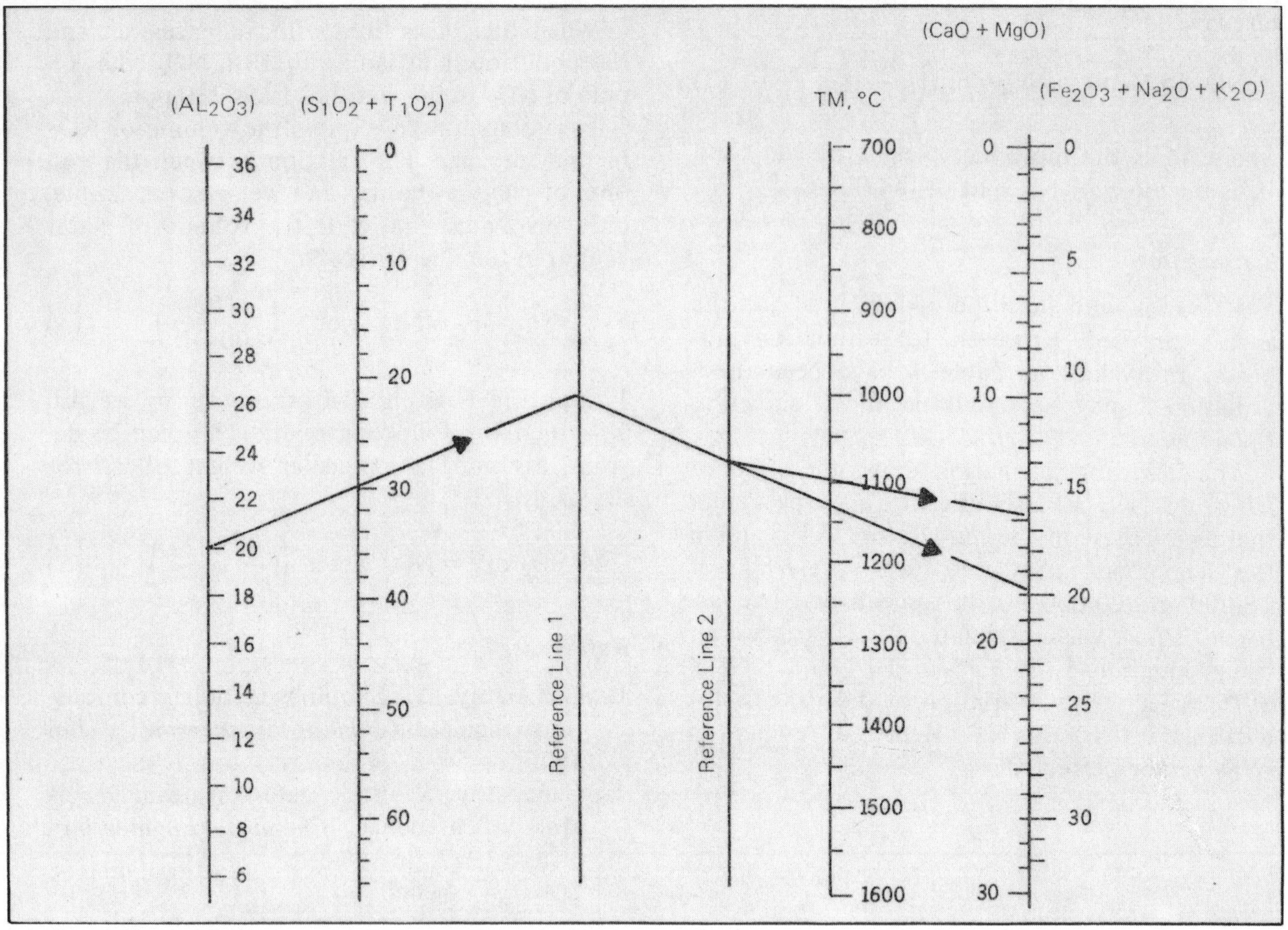


IN THE DESIGN OF STEAM GENERATORS AND ash removal systems it is important to consider the characteristics of the ash formed by burning the coal that will be used. The fusion temperature of the ash is a characteristic that has one of the greatest effects on design, and it is useful to be able to make a quick estimate of this temperature. An equation for this purpose which compares well with experimental data is as follows (from *Boiler House and Power Station Chemistry,* by W. Francis, 4th edition, p. 86, Edward Arnold Ltd):

$$T_M = 19\,Al_2O_3 + 15(SiO_2 + TiO_2) + 10(CaO + MgO) + 6(Fe_2O_3 + Na_2O + K_2O)$$

where $T_M$ = fusion temperature, $C$

All constituents are in percent by weight. The nomogram provides a quick estimate of the fusion temperature when the constituents are known.

***Example***

Analysis of a given ash indicates the following composition:

$A = Al_2O_3 = 20\%$
$B = SiO_2 + TiO_2 = 30\%$
$C = Fe_2O_3 + Na_2O + K_2O = 20\%$
$D = CaO + MgO = 15\%$

Find the fusion temperature.

***Solution***

Connect $A = 20$ with $B = 30$ and extend to cut Reference Line 1. Connect this point with $C = 20$ and mark point of intersection with Reference Line 2. Connect this point with $D = 15$ and read fusion temperature as 1100 $C$ where line crosses $T_M$ scale.

# 14. Nomogram determines value of waste heat —

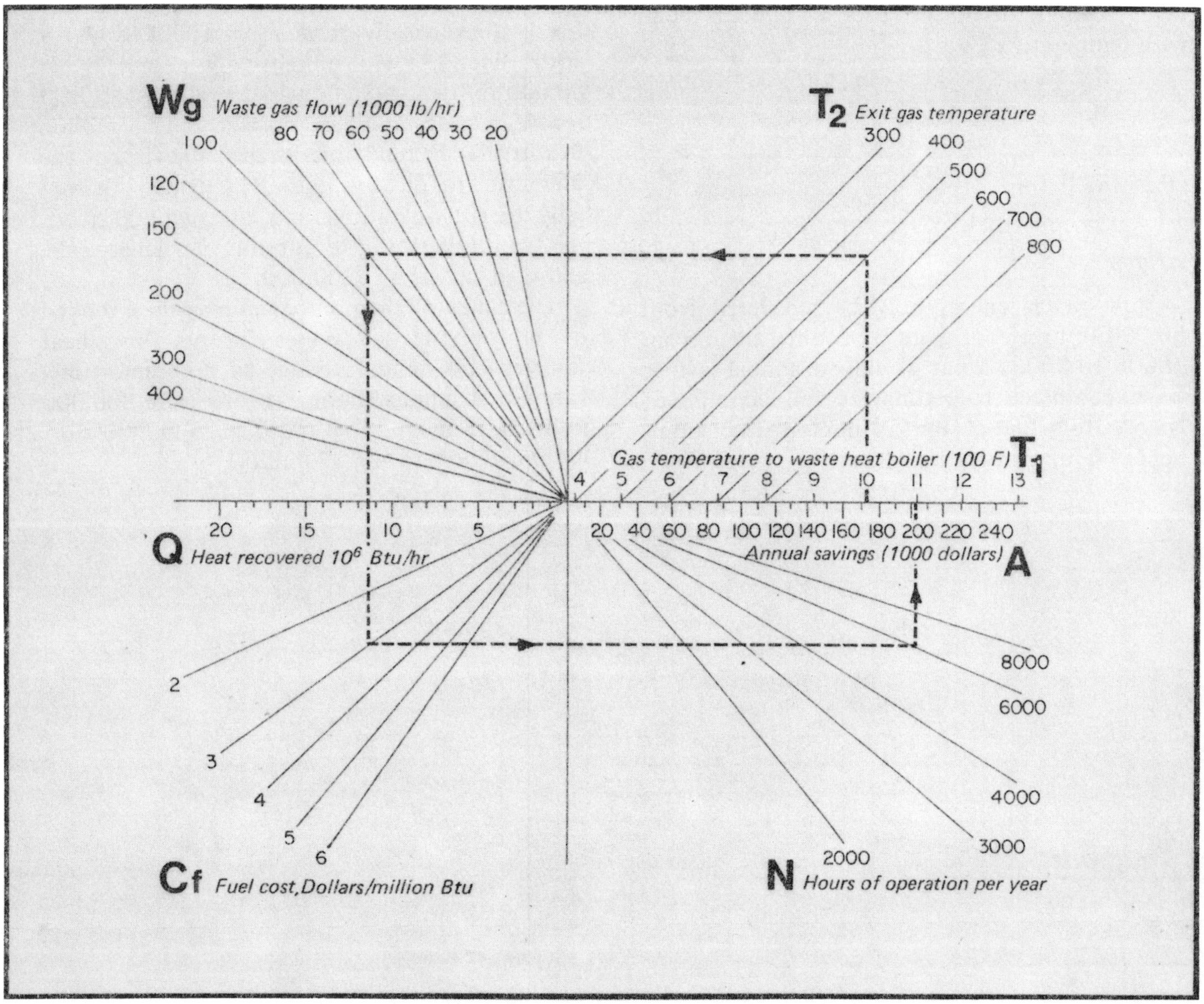


MANY PROCESSES PRODUCE LARGE VOLUMES of hot gases. As a first step to determining whether the energy in these gases should be recovered rather than letting it go to waste, and how it may be recovered, it is necessary to estimate the value of the energy that may be recovered. This nomogram permits such estimates to be made quickly.

The general relationships involved are as follows:

$$Q = W_g (h_1 - h_2)$$
$$A = QC_1N$$

where:

$Q$ = Heat recovered (Btu/hr)
$W_g$ = Quantity of hot gas available (lb/hr)
$h_1$ = enthalpy of hot gas
$h_2$ = enthalpy of gas at exit from heat recovery equipment
$A$ = Estimated savings due to heat recovery ($/yr)
$C_1$ = Cost of fuel ($/million Btu)
$N$ = Operating time of heat recovery equipment (hr/yr)

The nomogram solves these equations, and takes account of the variation of specific heat with temperature.

### *Example*

How much energy may be recovered from 80,000 lb/hr of hot gases by cooling them from 1000F to 500F? What would the annual savings be as compared to getting the same amount of energy from fuel costing $3/million Btu, assuming 6000 hr/yr operation?

### *Solution*

(1) Go vertically from $T_1$ = 1000F to $T_2$ = 500F line. (2) Go horizontally to cut $W_g$ = 80,000 lb/hr. (3) Go down vertically to cut heat recovery scale, and read answer as 11.5 million Btu/hr. (4) Extend this vertical line to cut the fuel cost line of $3/million Btu. (5) Go horizontally to cut operating time line of 6000 hr/yr. (6) Go up vertically to cut annual savings scale, and read answer as $200,000.

The range of the nomogram may be extended by multiplying the values for gas flow, heat transfer and annual savings by a common factor. For example, if the gas flow were 800,000 lb/hr, heat recovery would be 115 million Btu/hr, and annual saving $2 million.

# 15. Charts help estimate waste-heat recovery potential

IN SOME PROCESS PLANTS, ENERGY FROM various waste-gas streams and flue gases is not fully used. For energy-conscious plant engineers who must project to management the energy that could be recovered and the economics of it, two nomograms (Figs. 1 and 2) give assistance. They help estimate the following:

1. Energy recoverable from waste-gas streams and flue gases.
2. Amount of saturated steam that can be generated using a waste-heat boiler.
3. Annual fuel savings that can be accrued if a fired steam generator were installed instead of a waste-heat boiler.

Fig. 1 solves the following equations:

$$Q = W_g C_p (T_e - T_1) \quad (1)$$

$$W_s = Q/(H_s - H_{fw}) \quad (2)$$

where:

$Q$ = energy lost or recovered, Btu/hr
$W_g$ = gas quantity, lb/hr
$C_p$ = gas specific heat, Btu/lb-°F
$T_e$ = gas temperature entering, °F
$T_1$ = gas temperature leaving, °F
$W_s$ = steam generated, lb/hr
$H_s$ = enthalpy of saturated steam, Btu/lb
$H_{fw}$ = enthalpy of feedwater, Btu/lb
$T_{fw}$ = feedwater temperature, °F

An advantage of Fig. 1 is that no reference to steam tables is needed to calculate either $H_s$ or $H_{fw}$. The chart takes care of it as saturated-steam enthalpy at any pressure is 1,180 Btu/lb ±3%. Hence, a good estimate is obtained by taking $H_s$ as 1,180 Btu/lb.

Also, feedwater temperature $T_{fw}$, which is commonly known, is included in Fig. 1 and considers the corresponding enthalpy.

Gas specific heat $C_p$ is assumed to be 0.25 Btu/lb-°F.

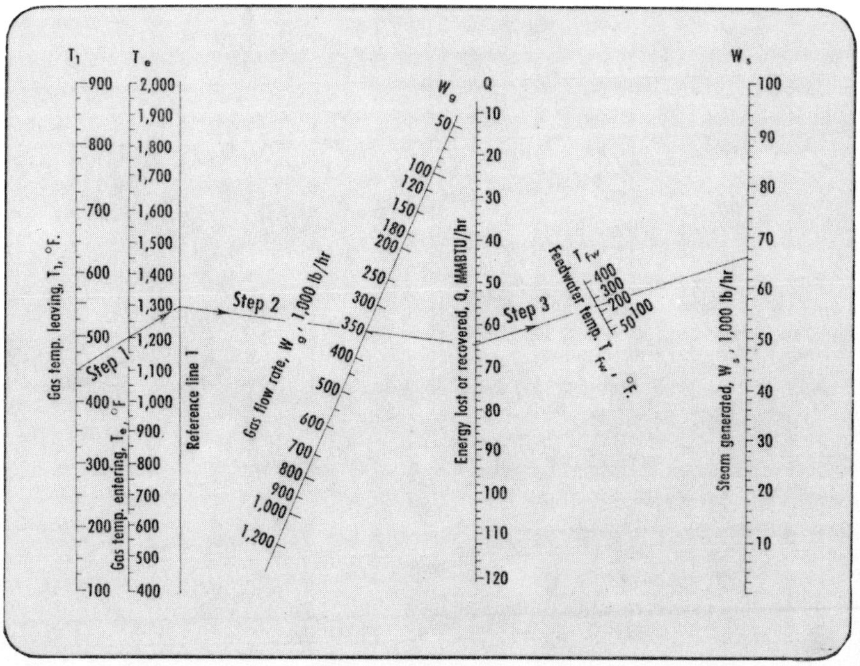


**Fig. 1.**

**Energy Available, Steam Generated**

Fig. 2 solves the following equation:

$$A = QCN/E \qquad (3)$$

where:

$A$ = annual savings in fuel bill, \$
$Q$ = energy lost or recovered, Btu/hr
$C$ = cost of fuel, \$/MMBtu
$N$ = annual operating period, hr
$E$ = steam generator efficiency, %

### *Example*

In a process plant, 350,000 lb/hr of flue gases are fed to the stack at a temperature of 1,200°F. It is felt that a waste-heat boiler could be installed to reduce the gas temperature to 450°F and to generate saturated steam at 600 psia with feedwater at 250°F.

Estimate the following:

1. Energy wasted.
2. Steam that may be generated using the waste-heat boiler.
3. Annual savings in fuel bill if a fired steam generator were used to generate steam at the same conditions with an efficiency of 80%. Fuel cost = \$1.8/MMBtu and the plant operates 6,000 hr/yr.

### *Solution*

In Fig. 1, connect $T_1$ = 450 with $T_e$ = 1,200 and extend to cut reference line 1 (Step 1). Connect that point with $W_g$ = 350,000 to cut the $Q$ scale and read 65 MMBtu/hr (Step 2). Connect $Q$ = 65 with $T_{fw}$ = 250°F and extend to cut the $W_s$ scale at 67,000 lb/hr (Step 3).

In Fig. 2, connect $Q$ = 65 MMBtu/hr with $E$ = 80% and extend to cut reference line 1 (Step 1). Connect that point with $C$ = \$1.8/MMBtu and extend to cut reference line 2 (Step 2). Connect that point with $N$ = 6,000 hr and extend to cut the $A$ scale (Step 3). Read \$870,000 as the annual savings in fuel cost.

Note that in Fig. 1, scales $W_g$, $Q$, and $W_s$ can have the same multiplication factor. Hence, if gas flow rate $W_g$ = 35,000 lb/hr, $Q$ would be 6.5 MMBtu/hr and $W_s$ would be 6,700 lb/hr.

If the cost of the waste-heat boiler, including controls and auxiliaries, was \$800,000, the annual savings, assuming a life of 10 years, is:

870,000−80,000 = \$790,000

Thus the approximate payback period, without considering interest charges, equals:

800,000/790,000 = 1 year

Thus a waste-heat-recovery boiler, instead of a fired steam generator, is a good proposition.

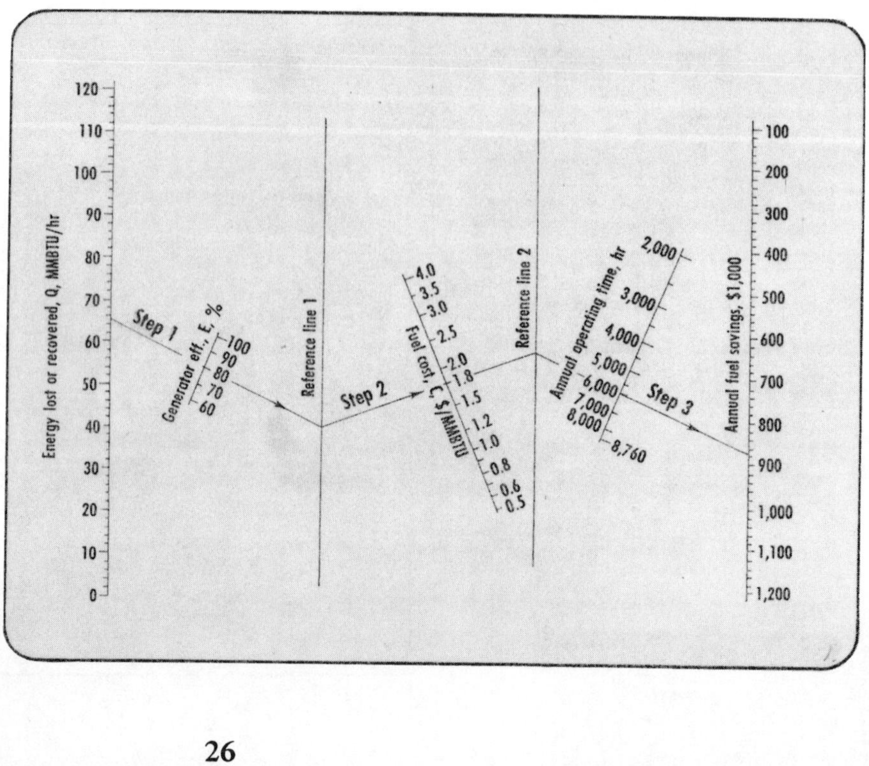


**Fig. 2. Annual Fuel Savings**

# Chapter 2

# Steam Generation and Usage

# 1. Nomograms provide superheated-steam properties at a glance

WHEN ENGINEERS NEED TO KNOW THE VALUES of superheated-steam properties, they generally go to the steam tables. However, frequent double-interpolation between temperatures and pressures can be cumbersome.

These two nomographs provide enthalpy and specific volume for superheated steam over the entire range of the steam tables. While the results are only approximations, they are accurate enough for many engineering applications.

*Examples*

1. What is the specific volume of superheated steam at 650 psia and 800°F?

2. What is the enthalpy of superheated steam at 650 psia and 800°F? If this steam is throttled to 300 psia, what is the final temperature?

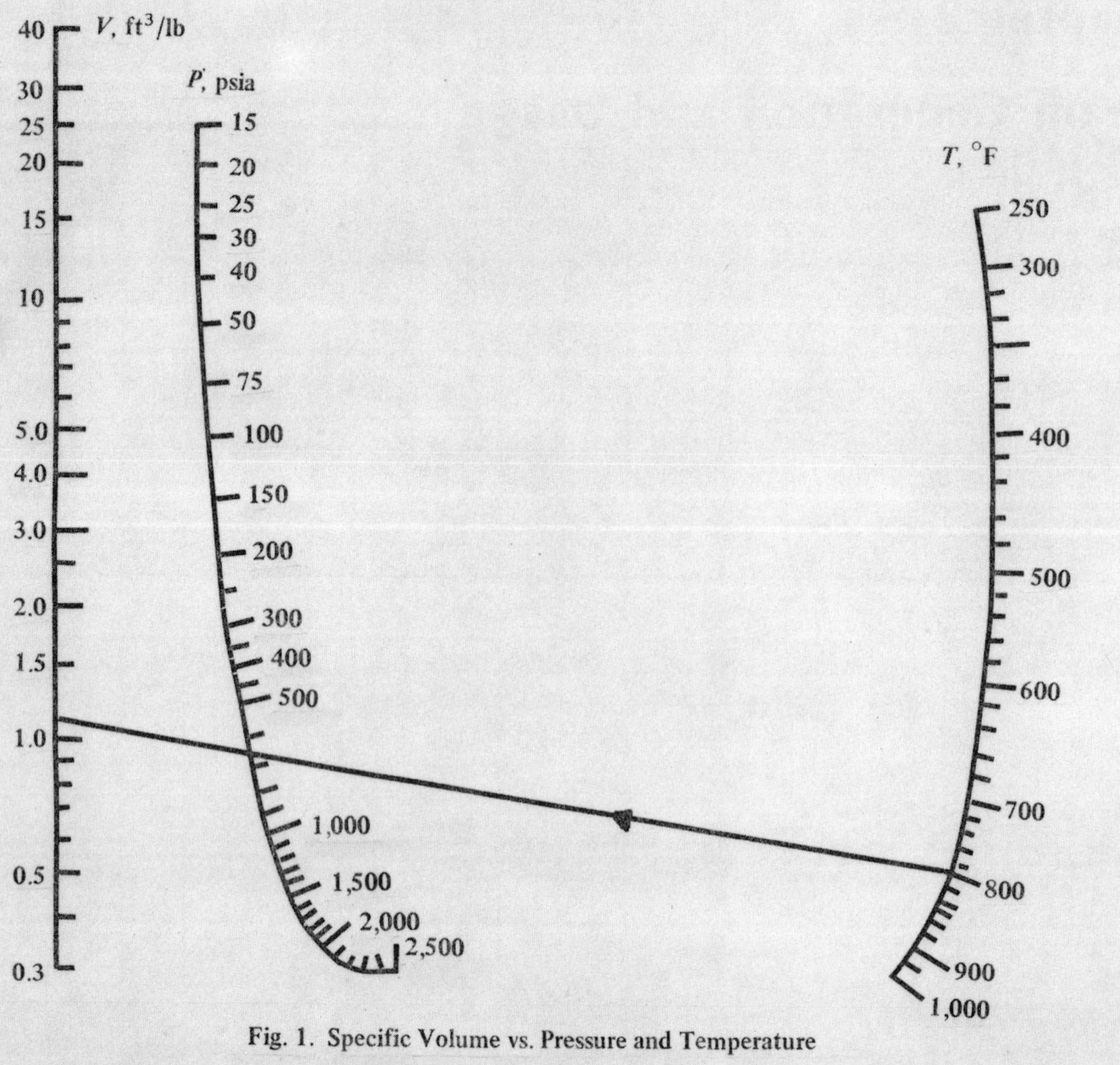


Fig. 1. Specific Volume vs. Pressure and Temperature

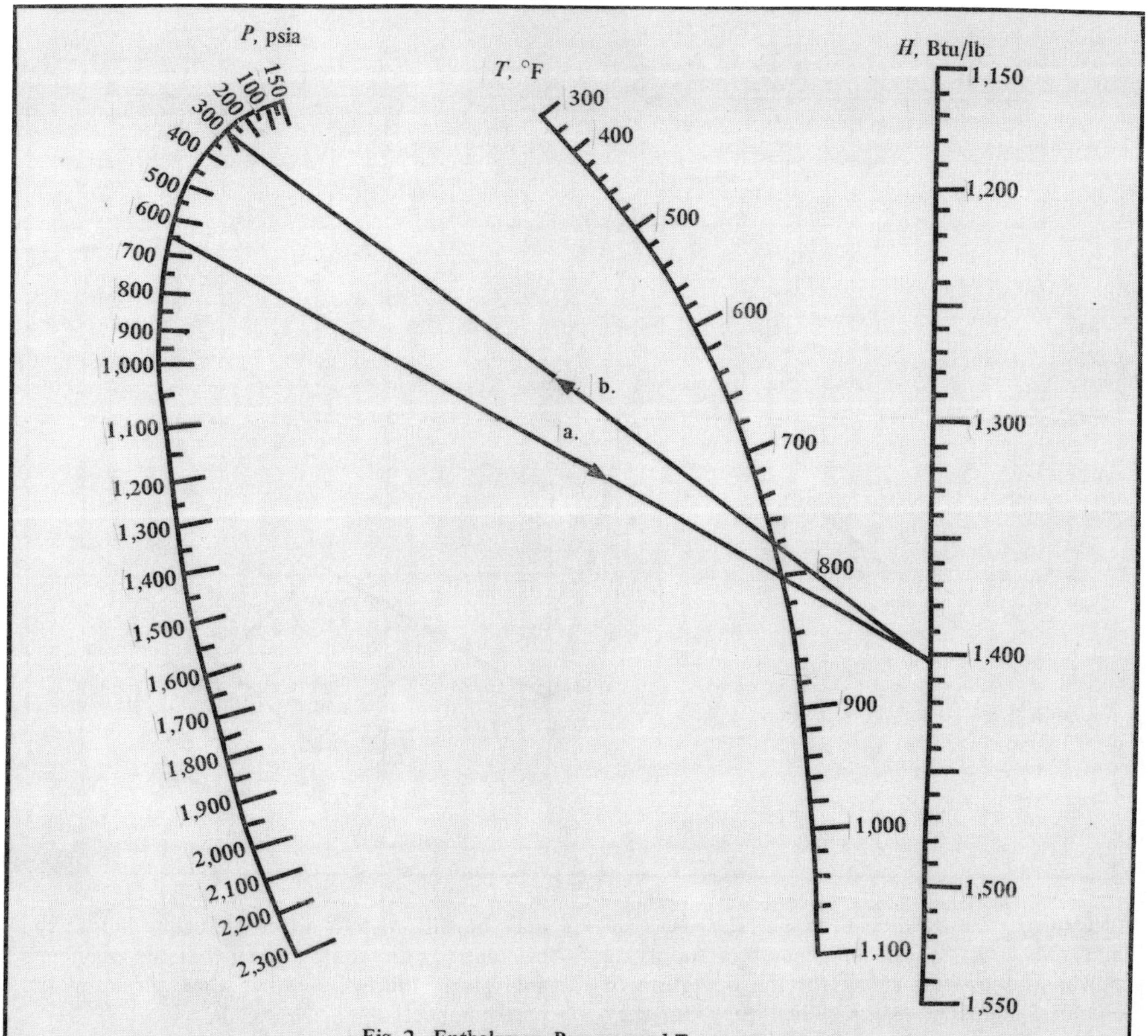


Fig. 2. Enthalpy vs. Pressure and Temperature

***Solutions***

1. On Fig. 1, connect $T = 800$ with $P = 650$, extend the line to the "$V$" scale and find the specific volume $V = 1.1$ ft$^3$/lb.

2. On Fig. 2, connect $P = 650$ with $T = 800$, extend the line to the "$H$" scale and read $H =$ 1,402 Btu/lb. Connect this point with $P = 300$ to obtain $T = 780$°F.

# 2. Charts provide basic data for steam generation at a glance

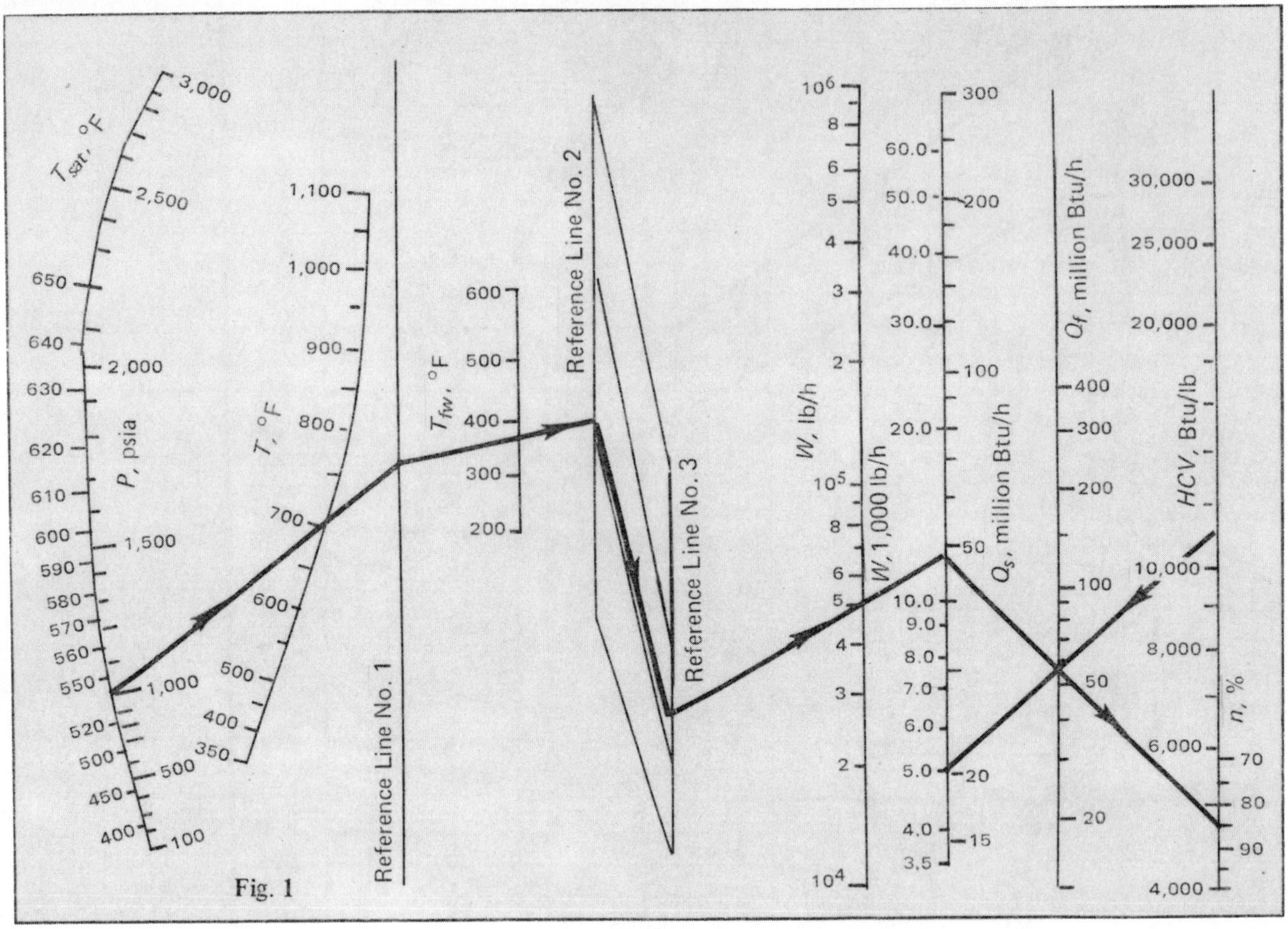


Fig. 1

CHEMICAL ENGINEERS FREQUENTLY BECOME involved with the steam generators supplying power and process steam for the operation of chemical industry plants. When they do, they need to have an idea of the following parameters for any given system:

- heat absorbed by the water and steam
- heat liberated by the combustion
- amount of fuel fired
- amount of combustion air
- combustion-air volume at operating temperature

These parameters can all be calculated by means of stoichiometry and heat balances, using steam tables and combustion data. However, the calculations are so involved that the furnace may be shifted to a different regime before the calculations are complete, so that the engineer's analysis is following rather than directing the performance.

These two charts may help with such situations. They incorporate steam-table data, so that reference to the tables is not necessary. Also, data for various fuels is included to allow comparison of the effects of the fuel on combustion air. The charts are based on the following equations:

$$Q_s = W_s(H_s - H_{fw})$$
$$Q_f = Q_s/n$$
$$W_f = Q_f/HCV$$
$$W_a = CaQ_f/10^6$$
$$V_a = W_a(460 + T)/2{,}400$$

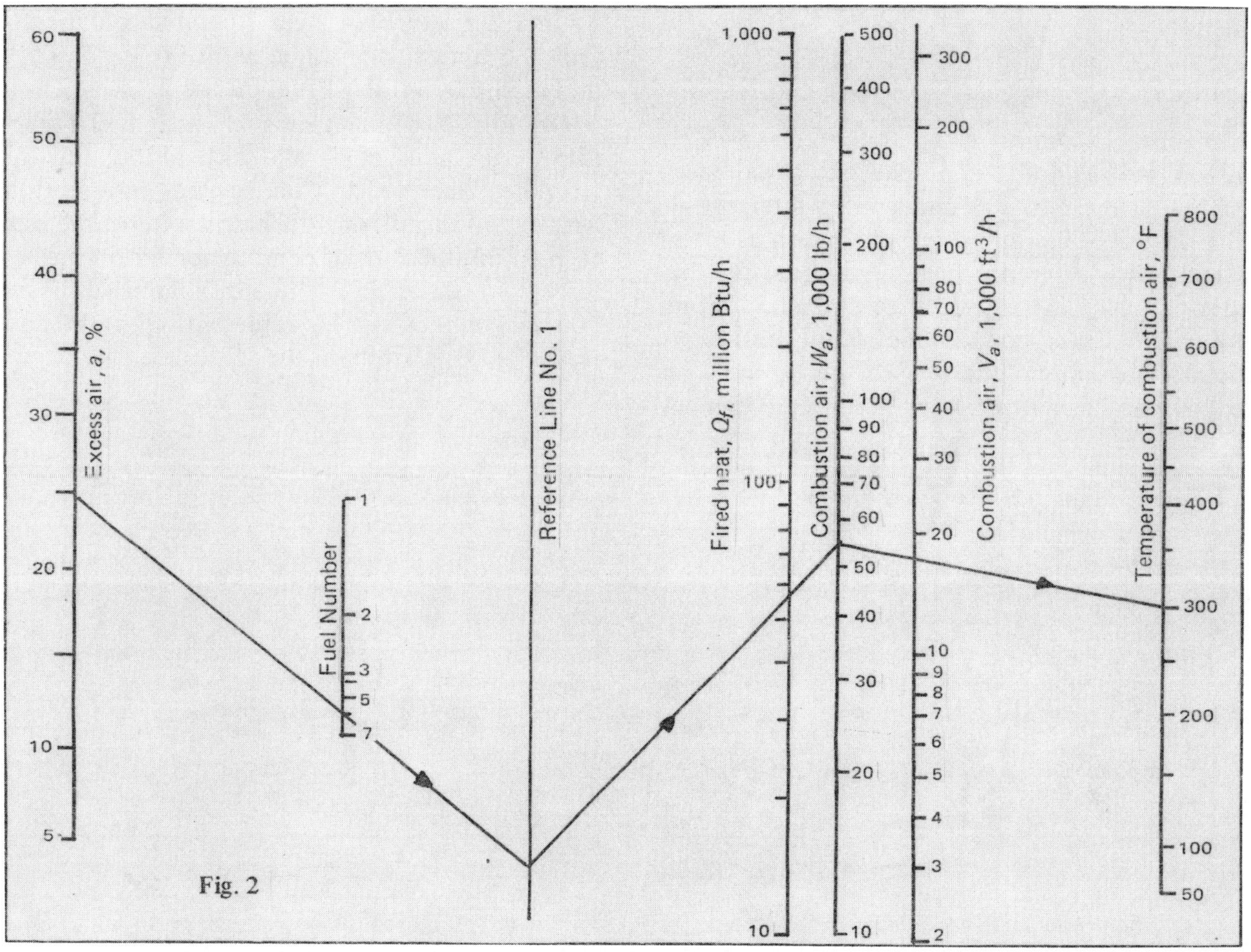


Fig. 2

where:

$Q_s$ = heat absorbed by the water and steam, Btu/h
$Q_f$ = heat fired, Btu/h
$W_s$ = steam generated, lb/h
$W_f$ = fuel fired, lb/h
$W_a$ = combustion air, lb/h
$V_a$ = combustion air, $ft^3$/min
$H_s$ = enthalpy of generated steam, Btu/lb
$H_{fw}$ = enthalpy of boiler feedwater, Btu/lb
$HCV$ = high heat of combustion of fuel, Btu/lb
$n$ = efficiency of boiler, %
$a$ = excess air factor (at 25% excess air, $a = 1.25$)
$C$ = a constant depending on the fuel
$T$ = temperature, °F

Values of $C$ for various fuels are as follows:

| *Type of fuel* | *Constant* | *Fuel No.* |
|---|---|---|
| blast-furnace gas | 570 | 1 |
| coke-oven gas | 670 | 2 |
| refinery gas and oil | 720 | 3 |
| natural gas | 730 | 4 |
| oil | 745 | 5 |
| bituminous coal | 760 | 6 |
| anthracite | 780 | 7 |

### *Example*

A boiler produces 50,000 lb/h of steam at 1,000 psia and 700°F from feedwater at 360°F by firing with 25% excess air a bituminous coal containing 10% ash and having a high calorific value of 11,000 Btu/lb. The boiler efficiency has been shown to be about 86%. What is the heat absorbed? the combustion heat liberated?

the amount of fuel fired? the amount of combustion air?

***Solution***

On Fig. 1, connect $P$ = 1,000 ($T_{sat}$ = 545°F) with $T$ = 700 and extend the line to intersect reference line No. 1. Connect this intersection with $T_{fw}$ = 360, and extend the line to its intersection with reference line No. 2. Transfer this intersection point to reference line No. 3 along the parallel lines, connect that point with $W_s$ = 50,000 and extend the line to the $Q_s$ scale to read 49 million Btu/h heat absorbed. Connect this value with 86% on the $n$-scale, and read 58 million Btu/h of heat liberated on the $Q_f$ scale. Connect this value with $HCV$=11,000, and read 5,000 lb/h of fuel on the $W_f$ scale.

On Fig. 2, connect $a$ = 25 with fuel No. 6 (for bituminous coal) and extend the line to reference line No. 1. Connect this intersection with $Q_s$ = 58 million Btu/h and extend the line to read 53,000 lb/h of combustion air on the $W_a$ scale. To determine the air volume at, say, 300°F, connect this $W_a$ value with $T$ = 300 and read 17,000 ft$^3$/min on the $V_a$ scale.

# 3. Nomogram estimates steam cost

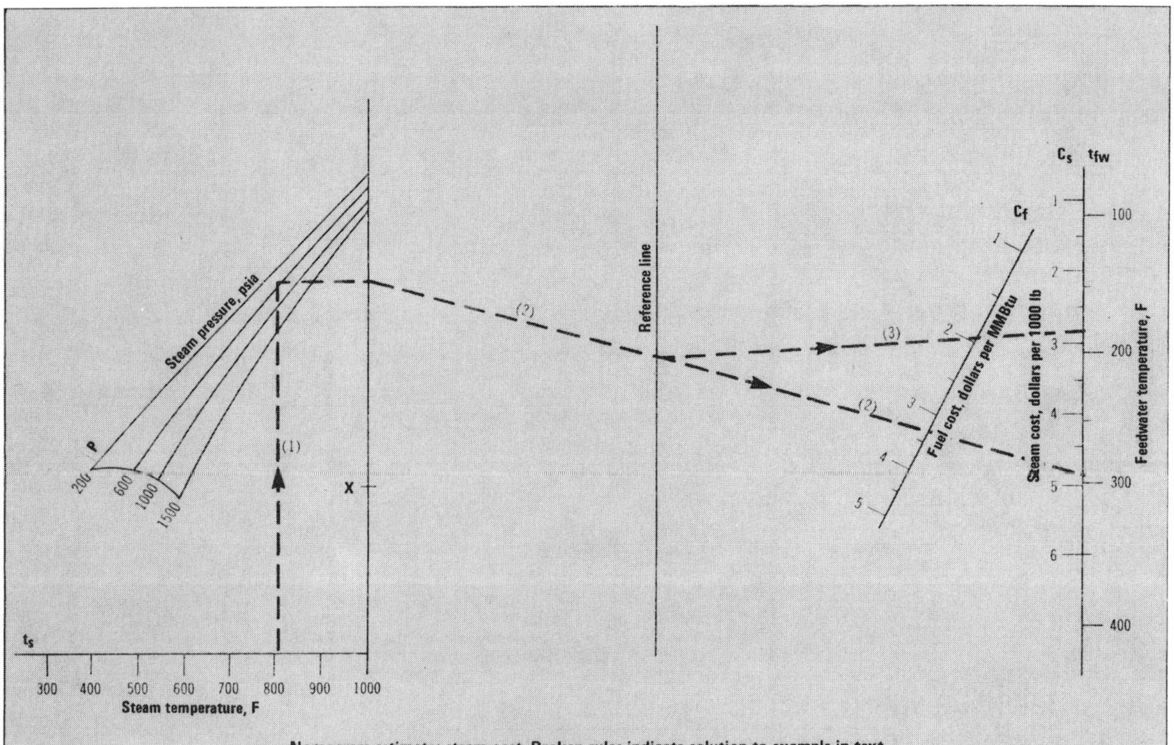


Nomogram estimates steam cost. Broken rules indicate solution to example in text.

A QUICK ESTIMATE OF STEAM COST CAN BE found with this nomogram if fuel cost, steam temperature, pressure, and feedwater temperature entering the boiler are known. It is not necessary to refer to steam tables. The efficiency of the boiler is assumed to be 80%.

The nomogram is based on the equation:

$$C_s = \Delta h \times 1000 \times C_f / \text{efficiency}$$

where:

$C_s$ = steam cost, dollars per MMBtu
$C_f$ = fuel cost, dollars per MMBtu
$\Delta h$ = enthalpy difference of $t_s$ at pressure $P$ and $t_{fw}$
$t_s$ = temperature of steam, F
$t_{fw}$ = temperature of feedwater

### *Example*

Estimate the cost of 1000 lb of steam if a boiler generates superheated steam at 800F and 600 psia with feedwater at 300F. The fuel cost is $2 per MMBtu.

### *Solution*

Draw a vertical line from 800 on the $t_s$ scale up to 600 psia on the $P$ scale. Connect this point with 300 on the $t_{fw}$ scale and mark where the line crosses the reference line. From this marked point, draw a line through 2 on the $C_f$ scale and extend it to the $C_s$ scale. Read $2.90 per 1000 lb as the solution.

Note that if you are working with saturated steam, begin at Point X and proceed as explained above. Also note, that a multiplication factor can be used to represent fuel costs higher than $5 per MMBtu. That same multiplication should be applied to the final answer. For example, if fuel cost was $8 per MMBtu (2X4), the solution would be 2.9X4, or $11.6 per 1000 lb.

# 4. Nomogram aids estimation of annual steam costs

THIS NOMOGRAM, FIG. 1, CAN be used to estimate annual steam costs. Fuel cost per 1 million Btu, steam generation rate, boiler efficiency, and operating hours must be determined before applying Fig. 1.

The nomogram is valid for saturated steam pressures up to 1500 psia and feedwater temperatures of about 250F. If superheated steam is required, Fig. 2 can be used to find the appropriate cost multiplication factor.

### *Problem*

What is the annual cost of generating 70,000 lb of steam per hour at 1000 psia if the fuel cost is \$1.20 per million Btu, boiler efficiency is 75%, and the unit operates 4000 hr?

### *Solution*

On the nomogram, align \$1.20 on scale *F* with 70,000 lb/hr on scale *S*, and extend the line to reference line 1. Connect this point with 75% on scale *E*. From the point at which reference line 2 is crossed, extend a line to 4000 hr on scale *B*, and, where scale *C* is intersected, read the annual cost of steam as \$450,000.

If superheated steam at 900F is required, the cost multiplication factor is 1.24 (Fig. 2). The annual cost for superheated steam is \$450,000 X 1.24 = \$558,000.

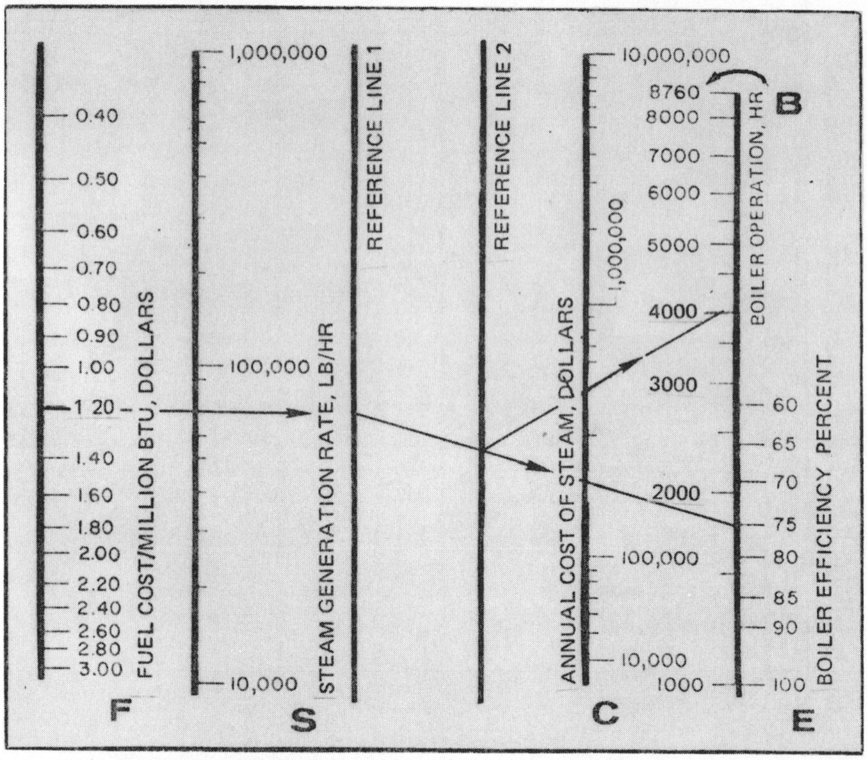


**Fig. 1. Nomogram can be used to estimate annual steam costs for a boiler operating at saturated steam pressures up to 1500 psia and with feedwater temperatures of about 250F.**

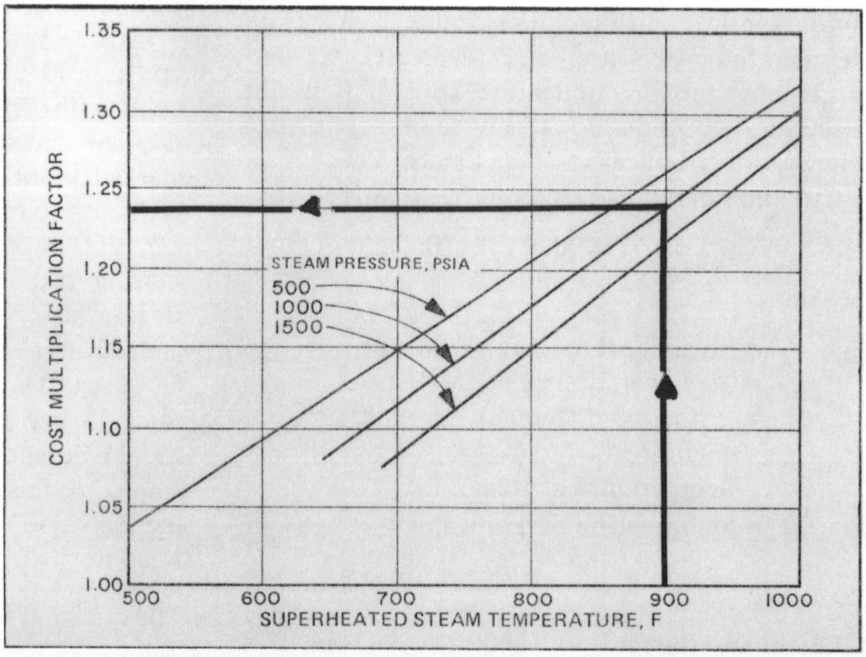


**Fig. 2. If superheated steam is required, a multiplication factor must be applied to the annual steam cost determined using Fig. 1. For example, if saturated steam at 900F is required from a boiler operating at 1000 psia, the annual cost figure calculated in Fig. 1 must be multiplied by 1.24.**

# 5. Chart estimates flash steam produced from blowdown and annual savings

BOILER BLOWDOWN CONTAINS A GREAT DEAL of heat, which can often be recovered as flash steam at a reduced pressure. This nomogram enables the plant engineer to calculate the quantity of flash steam that is generated and then can be used for process or space heating applications, and the annual savings that result.

### *Problem*

A 600 psia boiler operates 6000 hr annually and discharges 4000 lb/hr of blowdown. An energy conservation program that converts the saturated water to flash steam at 100 psia is established. How much flash steam is generated, and what are the annual savings if fuel costs \$2/million Btu?

### *Solution*

On the nomogram, connect 600 psia on Scale $P_1$ with the 100 psia curve on Scale $P_2$. Align this point on Scale $P_2$ with the 4000 lb/hr curve on Scale B and, where this line crosses Scale F, read the amount of flash steam as 19.5%. Therefore, 0.195×4000, or 780 lb/hr of flash steam is generated.

Connect the point on the 4000 lb/hr curve with the \$2/million Btu curve on Scale C; then move to the 6000 hr curve on Scale H. Connect the point on the 6000 hr curve with Scale S, and read the annual savings as \$11,000.

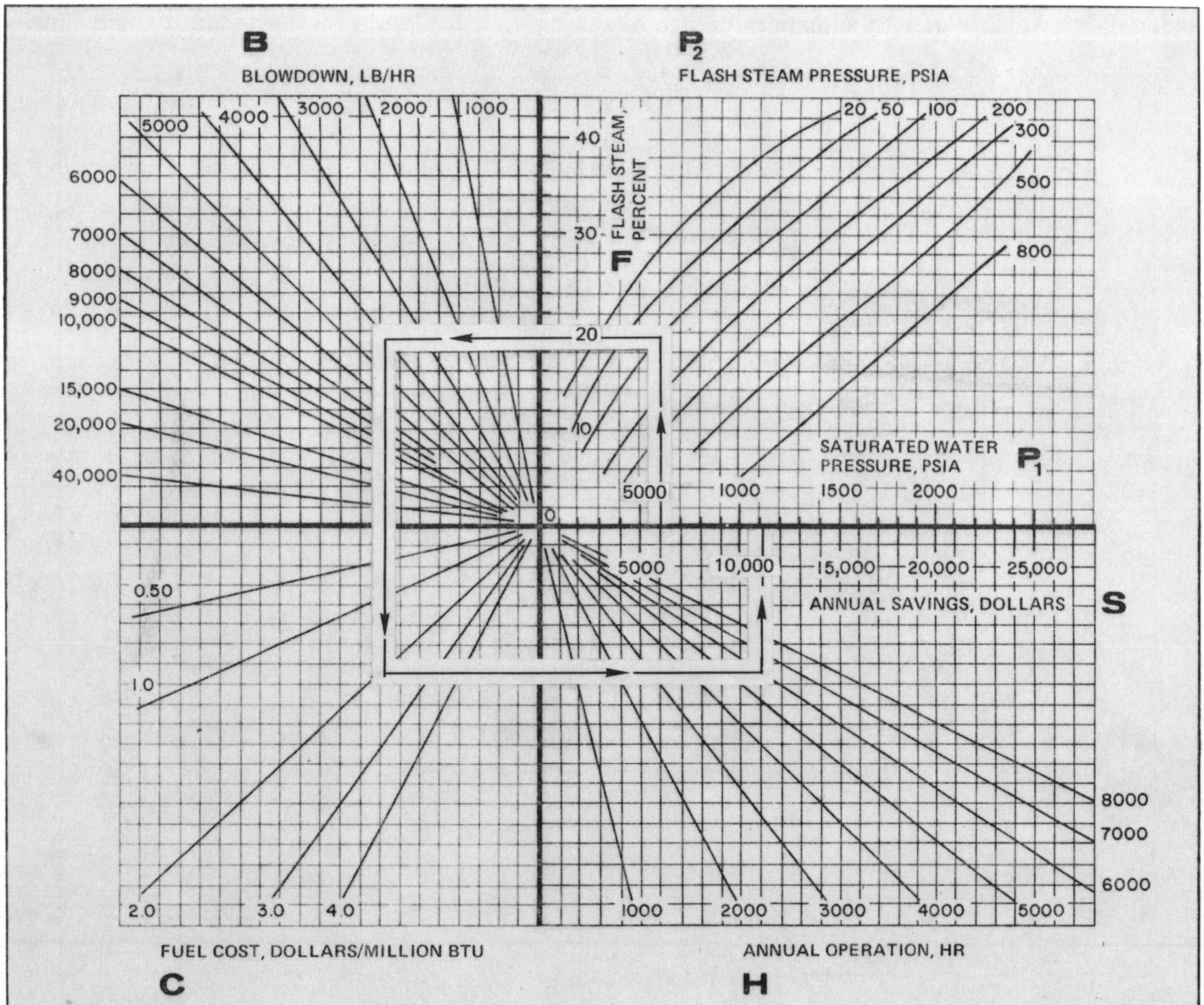

# 6. Chart estimates water required for steam-temperature control

ONE OF THE COMMONLY USED METHODS IN power and process plants for reducing the temperature of steam is by injecting the desired quantity of water in a desuperheater.

The water evaporates, in this way bringing down the final steam temperature.

To size the control valve and the lines, it is desirable to have an estimate of the spray water requirements, for various steam parameters. This may be obtained by performing an energy balance around the desuperheater.

$$(W_s - W)h_1 + Wh_{fw} = W_s h_2 \tag{1}$$

$W_s$ and $W$ refer to the final steam and spray water flow; $h_1$, $h_2$ refer to enthalpies of steam and the inlet and exit of the desuperheater. The term $h_{fw}$ refers to the enthalpy of the spray water.

Solving for $W$:

$$\frac{W}{W_s} = \left(\frac{h_1 - h_2}{h_1 - h_{fw}}\right) \tag{2}$$

Though this equation looks simple, the real problem lies in having to refer to steam tables for the enthalpies of steam and feedwater.

The accompanying chart avoids reference to steam tables. Its advantages are:

1. A lot of time is saved.
2. If the spray water quantity and steam

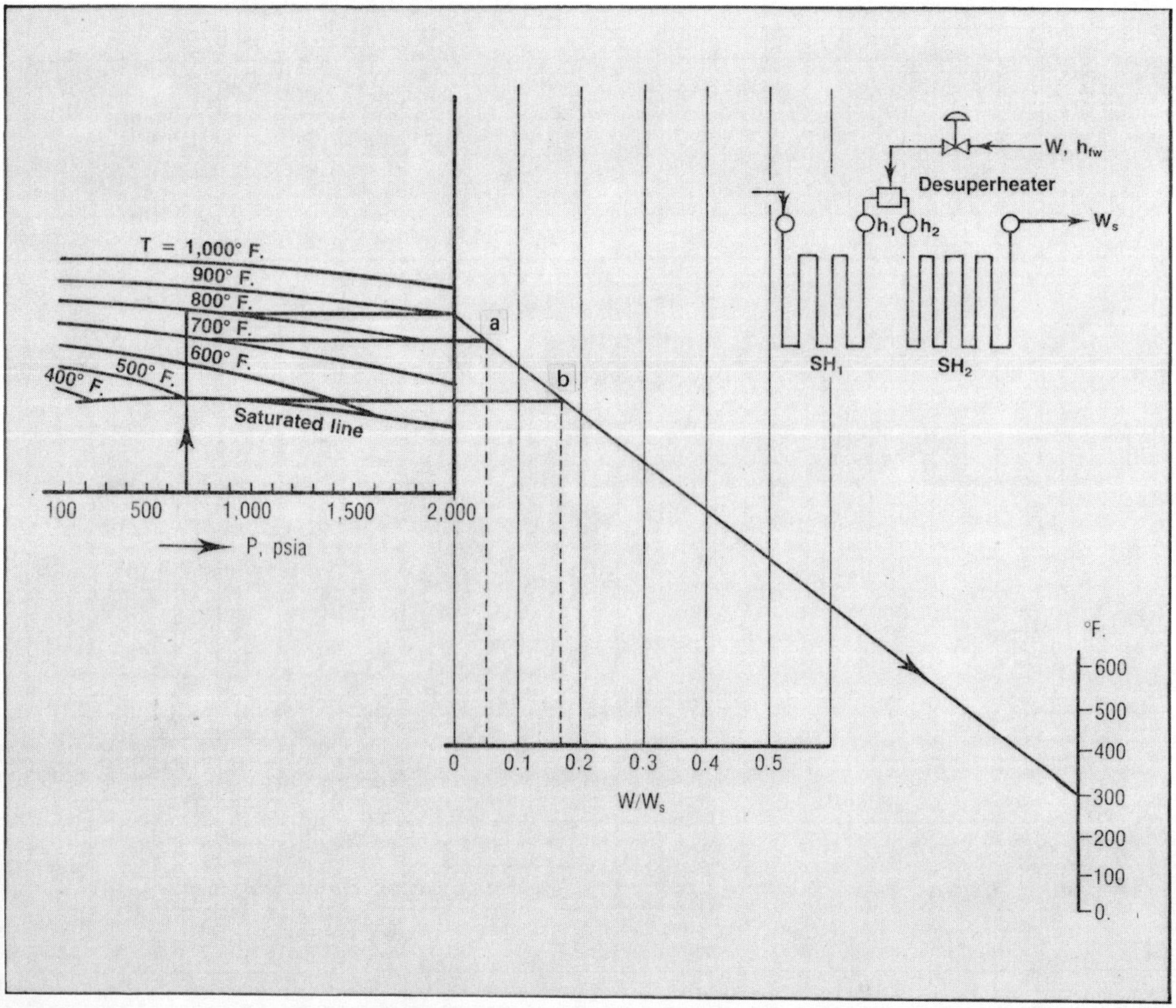

flow are known and if it is desired to estimate the final steam parameter, one has to resort to interpolation from steam tables data, thereby wasting time. With this chart, it is a simple affair.

Two examples illustrate the versatility of the chart.

### Example 1

Superheated steam at 700 psia and 800°F is to be reduced to saturated conditions using spray water at 300°F in a desuperheater.

Estimate the ratio $W/W_s$.

Neglect pressure drop in the desuperheater.

### Solution

Go up from $P = 700$ to cut the curve $T = 800°F$, move right to cut the reference line. Connect with spray water temperature of 300°F. Again, from the intersection of the vertical at $P = 700$ and the curve "saturated," move right to cut the earlier line at point b as shown.

Drop a vertical to read off $W/W_s = 0.17$. Hence 17% of the total steam is required as spray water.

### Example 2

If only 5% of the total flow is used for spray ($W/W_s = 0.05$), estimate the temperature to which the steam at 700 psia and 800°F is reduced.

### Solution

From $W/W_s = 0.05$, go up to cut the earlier line at point a. Move left to cut the vertical from $P$. The intersection occurs at $T = 700°F$, which is the final steam temperature.

Hence, the chart may be used for quick estimates of the performance of desuperheaters, without having to use the steam tables.

# 7. Nomogram relates steam volume and weight —

THE NOMOGRAM PRESENTED HERE CONVERTS steam by volume *(SBV)* to or from steam by weight *(SBW),* also known as dryness fraction or steam quality. It does this without the need to refer to steam tables to obtain the specific volumes of the liquid and vapor phases of the steam.

The nomogram is based on the equation,

$$SBV = 100/[1 + ([100 - X]/X)(V_w/V_s)]$$

where $SBV$ = steam by volume, %; $X$ = dryness fraction or steam by weight or quality, %; and $V_w$ and $V_s$ are the specific volumes of the saturated water and vapor phases, cu ft/lb. They are functions of the pressure of the steam.

If steam pressure and either $SBV$ or $X$ are known, they are sufficient to obtain the other variable with the nomogram. These variables are necessary to analyze circulation systems.

### *Example 1*

In a steam-water system at 1,000 psia, the dryness faction or quality *(X)* is 25%. What is the steam by volume?

### *Solution*

Connect $X$ = 25% on the nomogram with $P$ = 1,000 psia to cut the $SBV$ scale and read off 87%.

### *Example 2*

If steam by volume *(SBV)* and steam pressure are known, connect pressure *(P)* with $SBV$ and extend to cut the $X$ scale. Connecting $P$ = 1,000 psia (Example 1) with $SBV$ = 87% and extending to the $SBW$ scale gives $X$ = 25%.

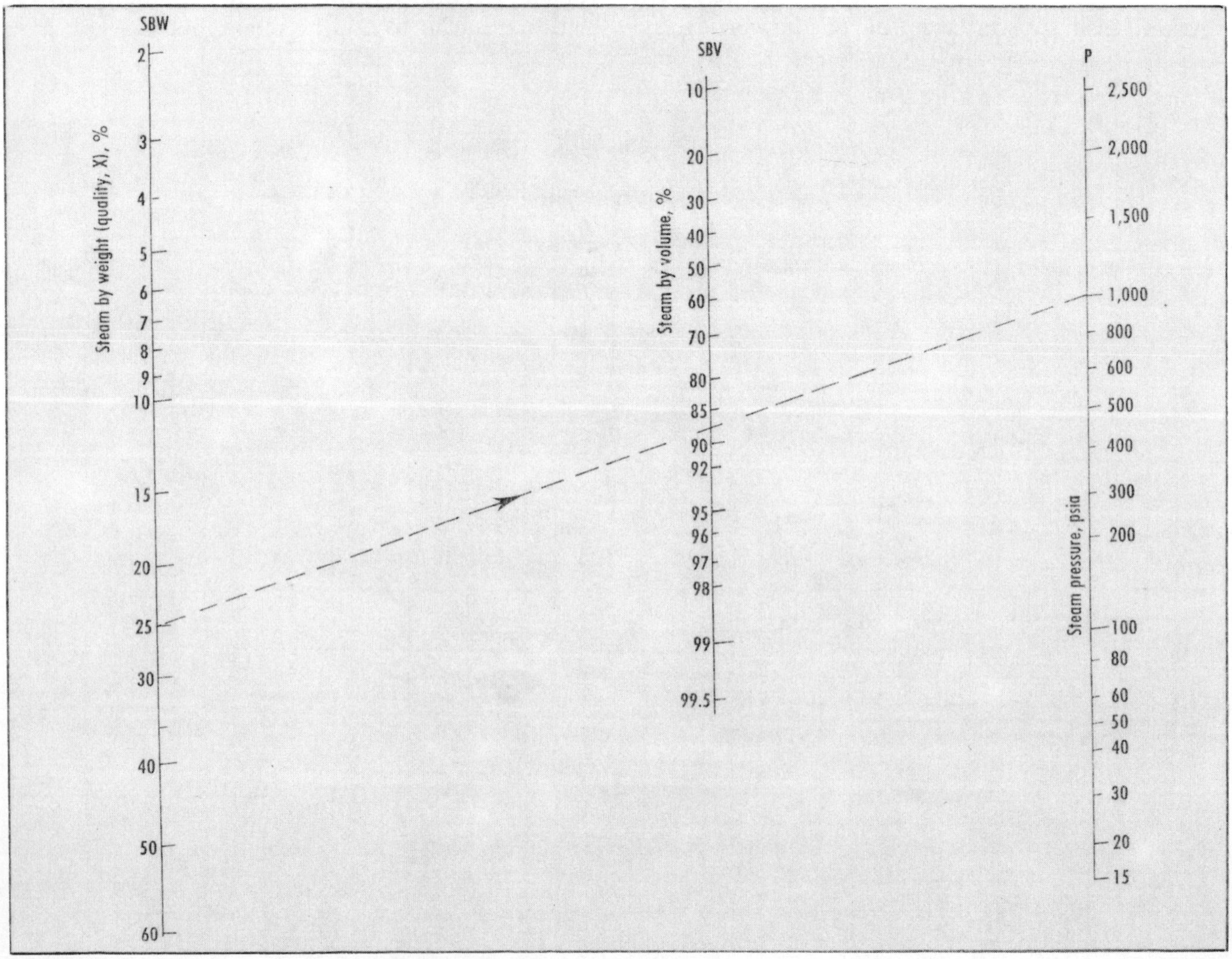

# 8. Nomogram determines volume of steam after throttling process

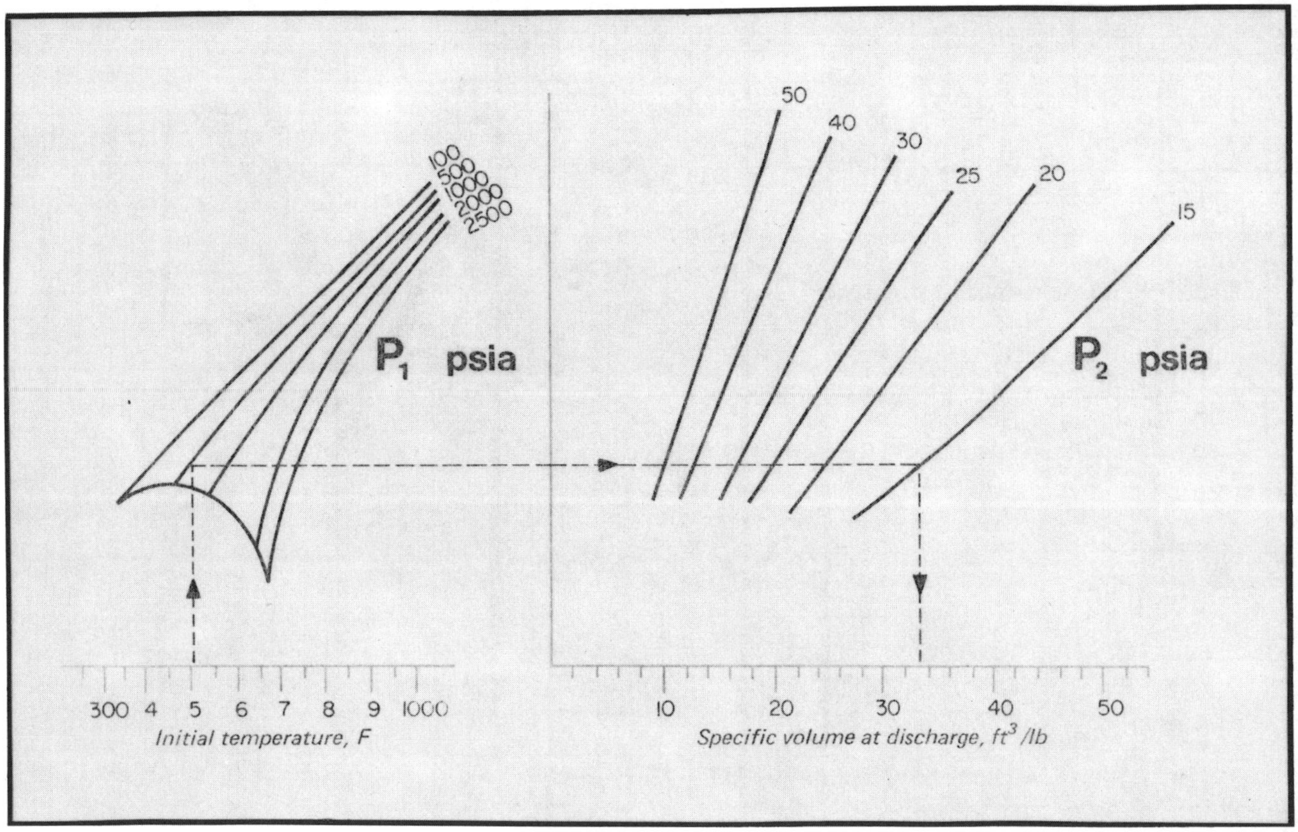


IN THE DESIGN OR SPECIFICATION OF SOME piping auxiliaries, such as silencers for steam vent lines, control valves, and blowoff lines, it is necessary to know the volume of steam after a throttling process. Assuming a constant enthalpy process, the calculation is not particularly difficult, but it does require the use of steam tables or a Mollier chart.

This nomogram determines the volume of steam after a constant enthalpy expansion without use of steam tables or Mollier chart. It is applicable for initially saturated or superheated steam.

***Example***

Superheated steam, initially at 500 psia and 500F, is blown off to atmospheric pressure at the rate of 60,000 lb/hr. Find the volume of steam at the discharge pressure.

***Solution***

(1) Go up vertically along 500F line to intersection with 500 psia initial pressure *($P_1$)*. (2) Go horizontally to intersection with discharge pressure line, 15 psia *($P_2$)*. (3) Go vertically down to intersect specific volume scale, and read 33 $ft^3$/lb.

Then, total discharge volume is equal to (33 X 60,000)/60 = 33,000 cfm.

# 9. Chart determines heater size and steam requirements for fuel oil preheaters

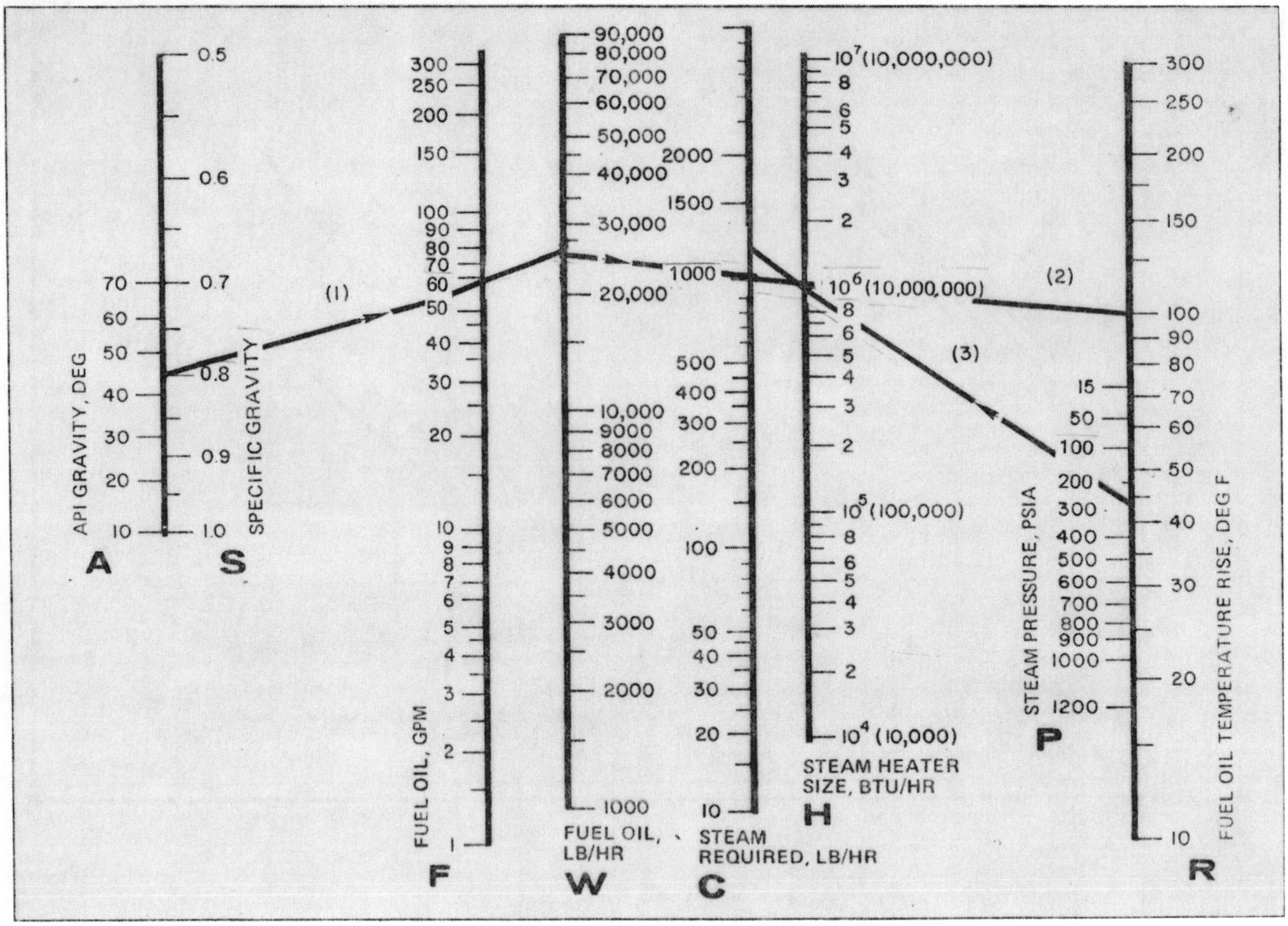


PREHEATING IS OFTEN NECESSARY TO FACILItate pumping and efficient burning of heavy oil (No. 5 and No. 6). Steam is one common method for preheating fuel oil.

This nomograph can be used to estimate the steam heater size and the amount of steam required. It is based on fuel oil with a specific heat of 0.43 Btu/lb/deg F, an accepted figure for engineering estimates. Steam tables do not have to be consulted.

### *Example*

Steam at 300 psia is used to preheat 60 gpm of fuel oil with a specific gravity of 0.8. What size of heater is needed and how much steam is required to raise the oil temperature from 50F to 150F?

### *Solution*

On the nomograph, align 0.8 on scale *S* with 60 gpm on scale *F* and extend the line to scale *W*. Connect the point of intersection on scale *W* (24,000 lb/hr) with 100F on scale *R*, and, where scale *H* is intersected, read heater capacity as slightly more than 1 million Btu/hr. Then connect this point on scale *H* with 300 psia on scale *P*, extend the line to scale *C*, and read the amount of steam required as 1250 lb/hr.

# 10. Nomogram provides steam consumption for air heating

QUICK ESTIMATES OF STEAM REQUIRED FOR air heating purposes may be found using the accompanying chart without the use of steam tables. Both British and metric units can be used.

Steam is used for heating air before it enters air heaters in boilers and fired heaters to avoid corrosion problems. To size the steam air heater, and associated control valves and lines, an estimate of heater duty and steam flow is needed.

The accompanying chart is based on the equation:

$$W_s = W_a \Delta T / LH = Q/LH$$

where:

$W_s$ = steam required, lb/hr or kg/hr
$W_a$ = air flow, lb/hr or kg/hr
$\Delta T$ = temperature rise, °F or °C
$C_p$ = specific heat of air, Btu/lb °F or kcal/kg °C

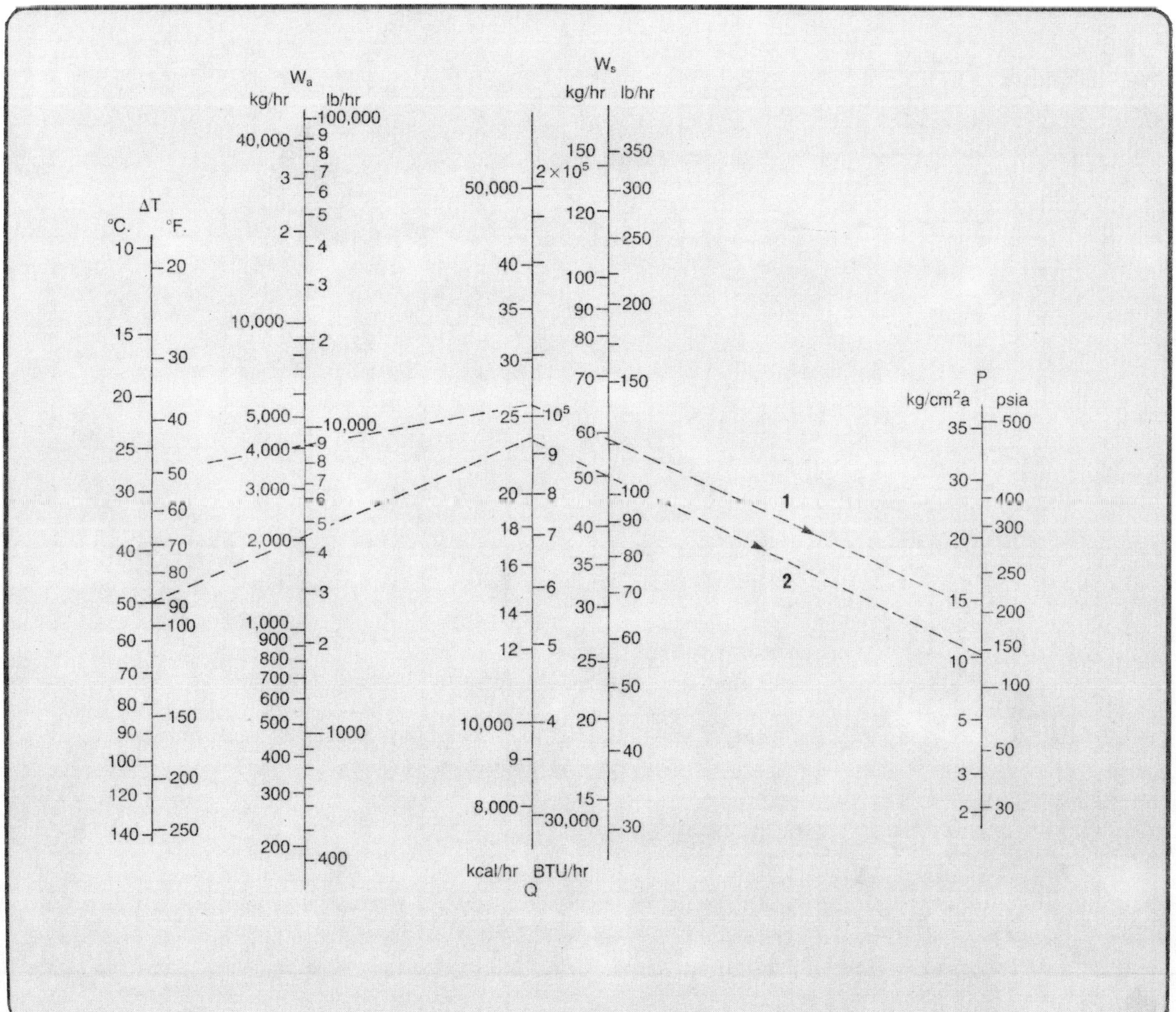

$Q$ = duty, Btu/hr or kcal/hr
$LH$ = latent heat of steam, Btu/lb or kcal/kg
$P$ = pressure, psia or kg/cm$^2$ a

*Example 1*

Assume 9,000 lb/hr of air must be heated through 50°F in a steam coil air heater using steam at 200 psia. What is the duty of the air heater and steam required?

For the solution, connect $\Delta T = 50$ with $W_a$ = 9,000 and extend to cut $Q$ scale. Read off 110,000 Btu/hr. Connect with $P$ = 200 to cut $W_s$ scale at 126 lb/hr.

*Example 2*

If 2,000 kg/hr of air must be heated through 50°C using steam at 10 kg/cm$^2$ a, what are $Q$ and $W_s$?

Following the same procedure as above, $Q$ = 24,000 kcal/hr and $W_s$ = 50 kg/hr.

# Chapter 3

# Heat Transfer

# 1. Nomograms determine inside heat transfer coefficients of common gases and liquids

INSIDE HEAT-TRANSFER COEFFICIENT OF THE gas or liquid flowing through heat-exchanger tubes is an important factor in calculating overall heat transfer or in estimating tube metal temperature.

The overall heat-transfer coefficient can be found from:

$$\frac{1}{U} = \frac{1}{H_i} + \frac{1}{H_o}$$

where: $U$ = overall heat transfer coefficient
$H_i$ = inside heat transfer coefficient
$H_o$ = outside heat transfer coefficient

In a heat exchanger with gases flowing inside the tubes and liquids flowing outside, the inside heat-transfer coefficient, which will be the smaller value, will be the governing factor in the efficiency of the heat exchanger.

Since the outside heat-transfer coefficient is generally known, or can be determined from reference materials, it is important to be able to compute the inside heat-transfer coefficient.

The four nomographs (on the next four pages) can be used to quickly determine the inside heat-transfer coefficients of superheated steam, common gases and liquids, and refrigerants flowing inside tubes.* For superheated steam, the gas or liquid mass flow rate and temperature, tube diameter, and pressure must be known. Data on specific heat, viscosity, thermal conductivity, gas or liquid velocity, and tube material and thickness are unnecessary for this calculation.

*Example 1*

What is the inside heat-transfer coefficient of 5000 lb/hr of 800F superheated steam flowing through a 1.5 in.-dia tube at 1600 psia?

***Solution***

On nomograph 1, connect 800F on scale *T* with 1600-psia curve on scale *P;* from this point, move horizontally to Reference Line 1. Connect the point on Reference Line 1 with 1.5 in. on scale *D* and extend the line to Reference Line 2. Align the point on Reference Line 2 with 5000 lb/hr on scale *G* and, where the line crosses scale *H,* read the inside heat-transfer coefficient as 430 Btu/(hr) (sq ft) (deg F).

*Example 2*

What is the inside heat-transfer coefficient of 400 lb/hr of 400F air flowing through a 2-in.-dia tube?

***Solution***

On Nomograph 2, connect 400F on scale *T* with Gas Curve 2, and move horizontally to Reference Line 1. Connect the point on Reference Line 1 with 2 in. on scale *D* and extend the line to Reference Line 2. Align the point on Reference Line 2 with 400 lb/hr on scale *G.* Where the line crosses scale *H,* read the inside heat-transfer coefficient as 15 Btu/(hr) (sq ft) (deg F).

*Example 3*

What is the inside heat-transfer coefficient of 2000 lb/hr of 300-F water flowing through a 0.75-in.-dia tube?

***Solution***

On Nomograph 3, connect 300F on scale *T* with liquid curve 10 and move horizontally to scale *D.* Connect this point with 2000 lb/hr on scale *G* and mark the point of intersection with the Reference Line. Align 0.75 in. on scale *D* with the marked point on the Reference Line, extend the line to scale *H,* and read the inside heat-transfer coefficient as 1600 Btu/(hr) (sq ft) (deg F).

*Example 4*

What is the inside heat-transfer coefficient of 200 lb/hr of 40-F R-114 refrigerant vapor flowing through a 1-in.-dia tube?

***Solution***

On Nomograph 4, connect 40F on scale *T* with Refrigerant Curve 5 and move horizontally to Reference Line 1. Connect this point with 200 lb/hr on scale *G* and extend the line to Reference Line 2. Align the point on Reference Line 2 with 1 in. on scale *D,* extend the line to scale *H,* and read the inside heat-transfer coefficient as 15 Btu/(hr) (sq ft) (deg F).

*Construction of the nomographs in this article is based on the relationship between Nusselts Number, Reynolds Number, and Prandtl's Number

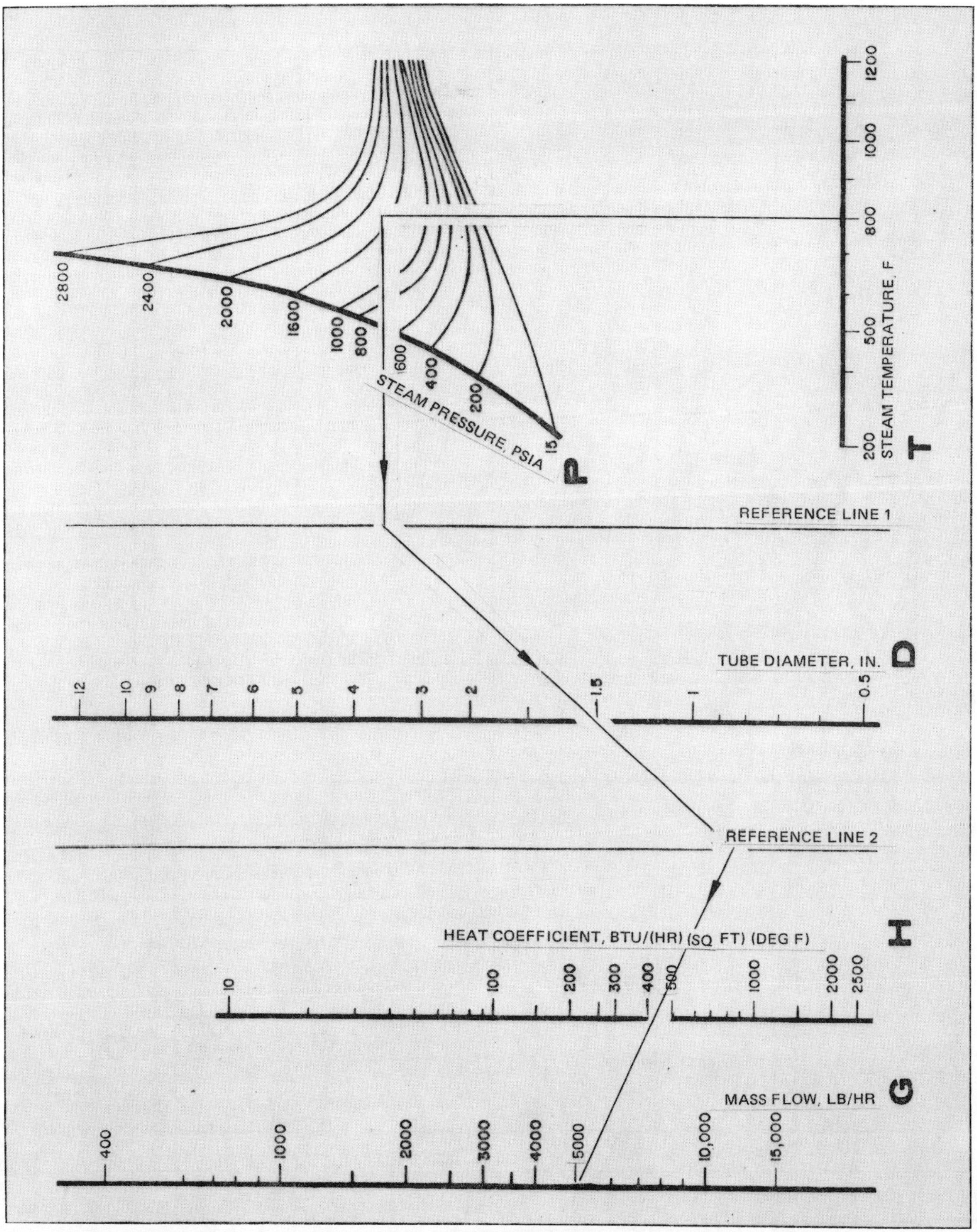


**Nomograph 1—for determining the inside heat transfer coefficient of superheated steam**

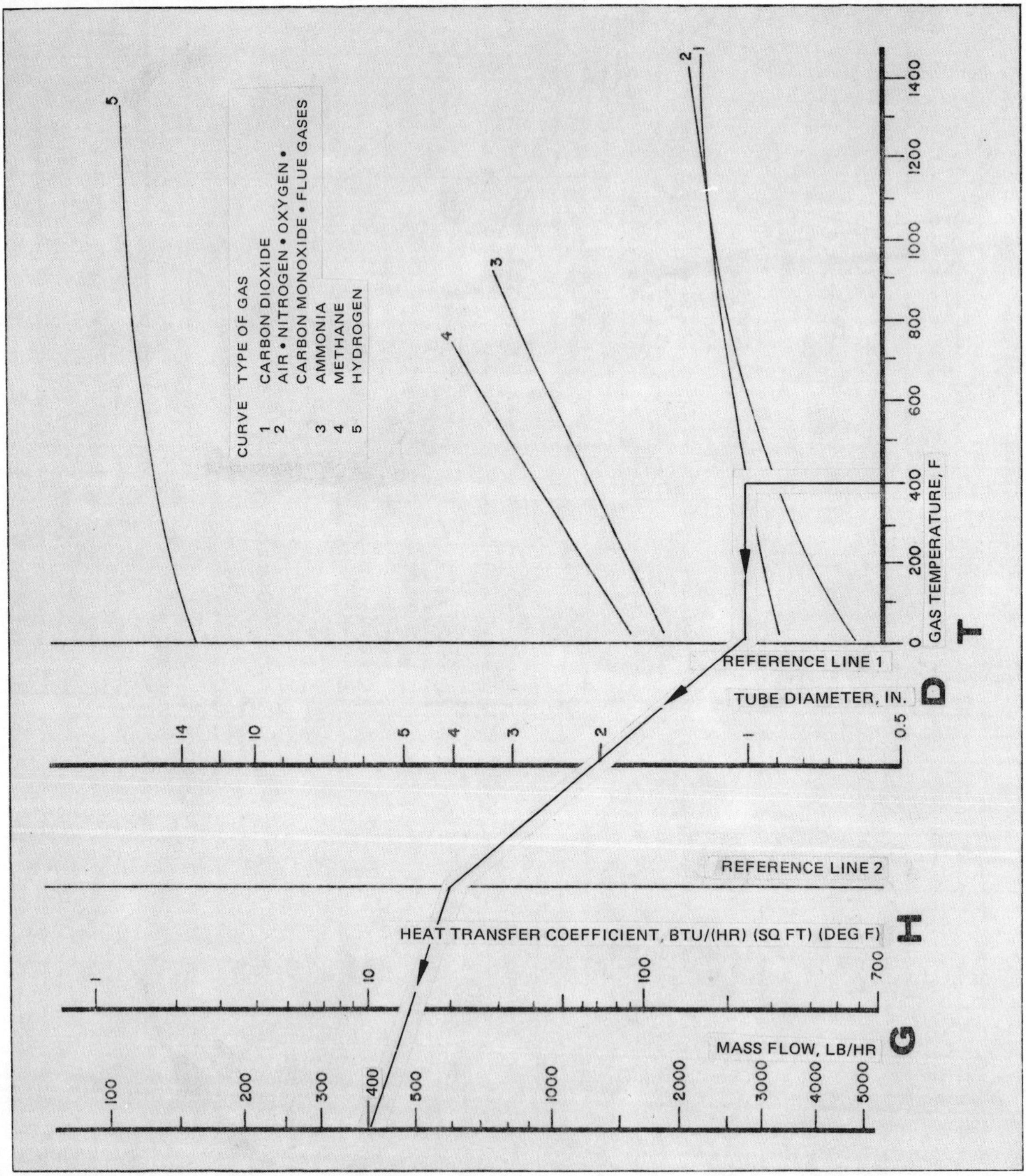


**Nomograph 2—for determining the inside heat transfer coefficient of common gases**

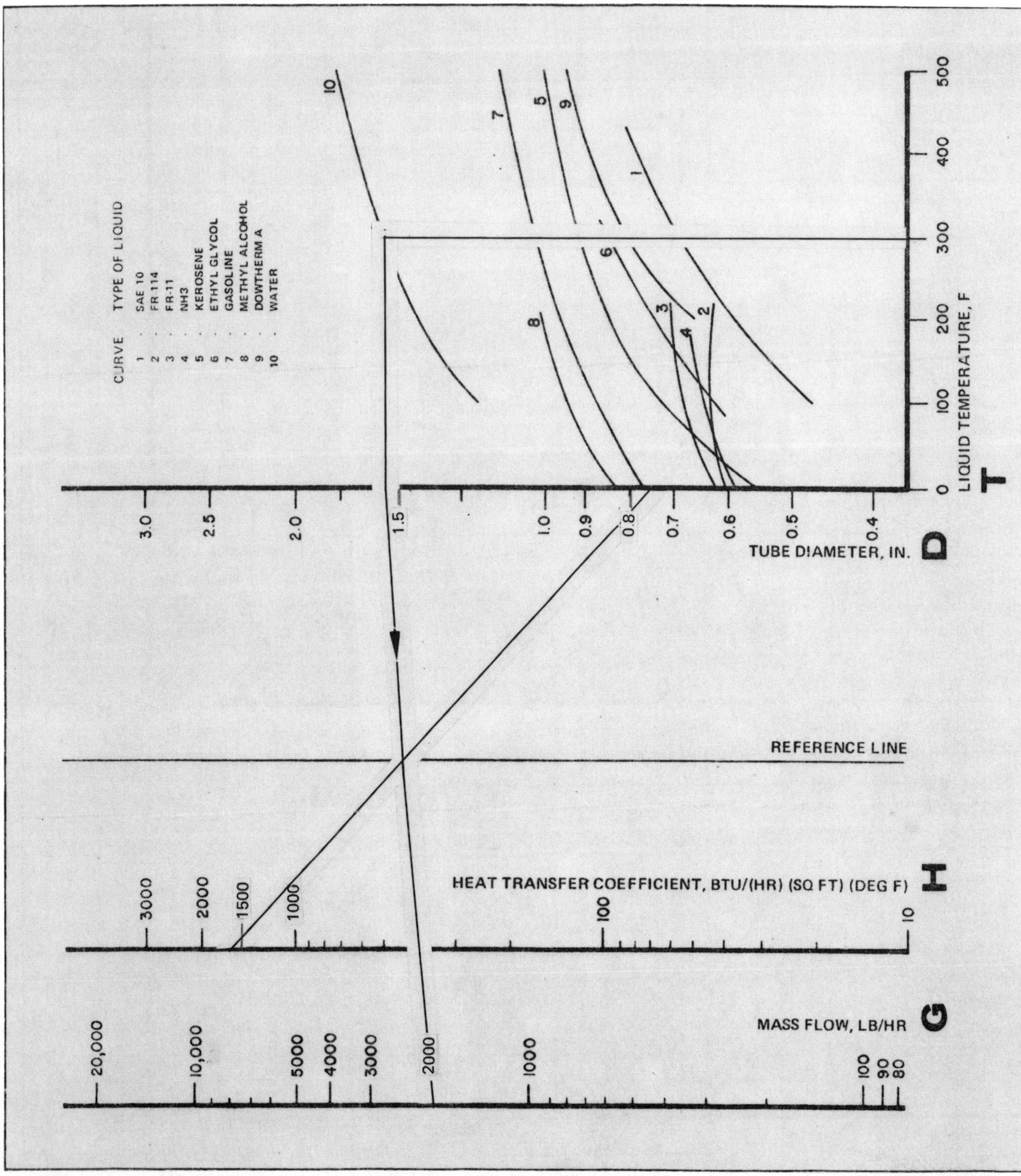


**Nomograph 3—for determining the inside heat transfer coefficient of common liquids**

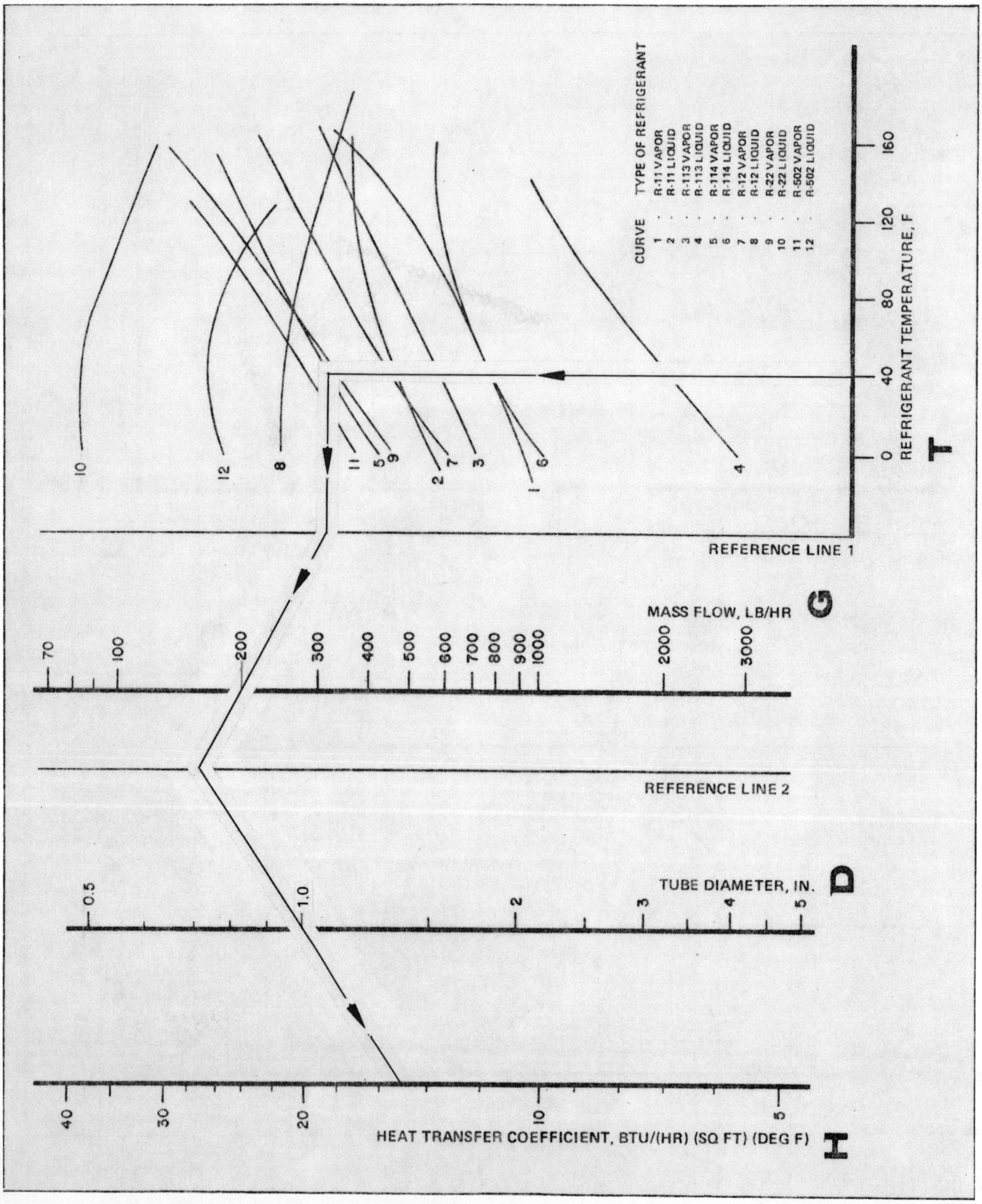


**Nomograph 4–for determining the inside heat transfer coefficient of several common refrigerants**

# 2. Chart gives fast estimate of gas heat-transfer coefficients

THE NOMOGRAM MAKES A QUICK WAY TO FIND shell-side, heat-transfer coefficients for gases in an exchanger with a triangular pitch arrangement. Its advantage is that no reference to thermal or transport properties is required. It is sufficient that the gas temperature *(T),* mass velocity *(G),* and tube size *(d)* are known.

The basic equation for the shell-side coefficient may be written as:

$$\text{Nu} = 0.36\ \text{Re}^{0.55}\,\text{Pr}^{0.33} \qquad (1)$$

This equation may be rewritten as:

$$H = 1.1\,(G^{0.55}/d^{0.45})\,C_p{}^{0.33}k^{0.67}/\mu^{0.22} \qquad (2)$$

where:

Nu = Nusselt's Number = $Hd/12k$
Re = Reynold's Number = $Gd/12\mu$
Pr = Prandtl's Number = $\mu C_p/k$
$H$ = gas heat transfer coefficient, Btu/ft²-hr-°F.
$d$ = tube OD, in.
$G$ = gas mass velocity, lb/ft²-hr
$\mu$ = gas viscosity, lb/ft-hr
$k$ = gas thermal conductivity, Btu/ft-hr-°F
$C_p$ = gas specific heat, Btu/lb-°F
$T$ = gas temperature, °F

Equation 2 is tedious to solve as the various terms are raised to complex powers, and reference to thermal or transport properties of gases may involve a lot of time. But with the nomogram, it is easy to get the heat-transfer coefficient.

Also, the term $F = C_p{}^{0.33}\ k^{0.67}/\mu^{0.22}$ is plotted on one of the scales so that if gases other than those mentioned are used, the term $F$ may be directly used. For other gases, the plot of $F$ vs. $T$ is already available.

### *Example*

Flue gases flow over a bundle of tubes of a heat exchanger. Estimate the gas heat-transfer

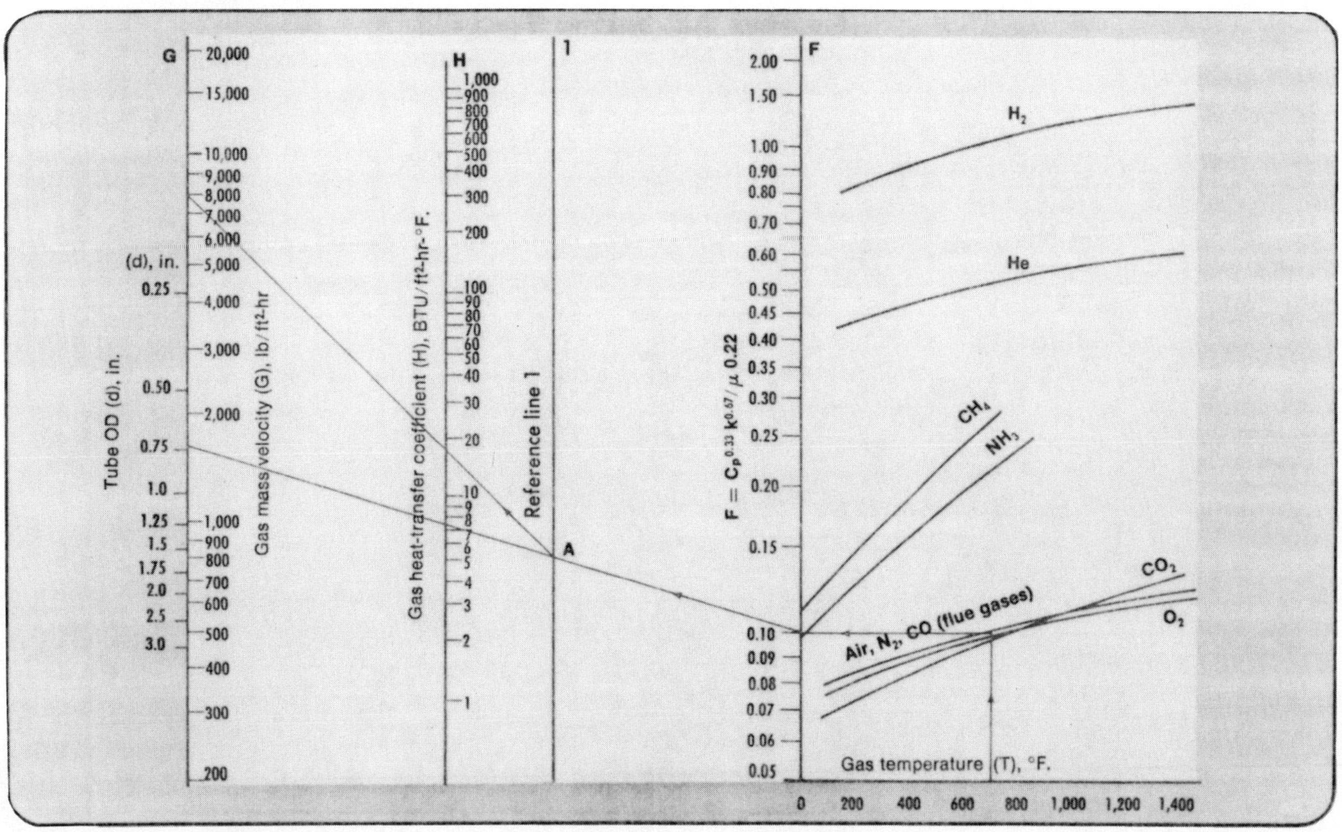

coefficient when $T$ = 700°F, $G$ = 8,000 lb/ft²-hr, and $d$ = 0.75 in.

### *Solution*

On the right side of the nomogram, go vertically from $T$ = 700 to cut the curve marked flue gases. Go horizontally left to cut the $F$ scale. Connect with $d$ = 0.75 to cut reference line 1 at $A$. Connect with $G$ = 8,000 to give $H$ = 18 Btu/ft²-hr-°F.

An added advantage of the chart is that the effect of $G$ and $d$ on $H$ can be quickly studied.

### *Bibliography*

1. Ganapathy, V., "Shell side heat transfer coefficients found fast for liquids," *OGJ,* Oct. 30, 1978, p. 114.
2. Ganapathy, V., "Quick estimation of gas heat-transfer coefficients," *Chemical Engineering,* Sept. 13, 1976, p. 199.

# 3. Charts provide shell-side heat-transfer coefficients for liquids

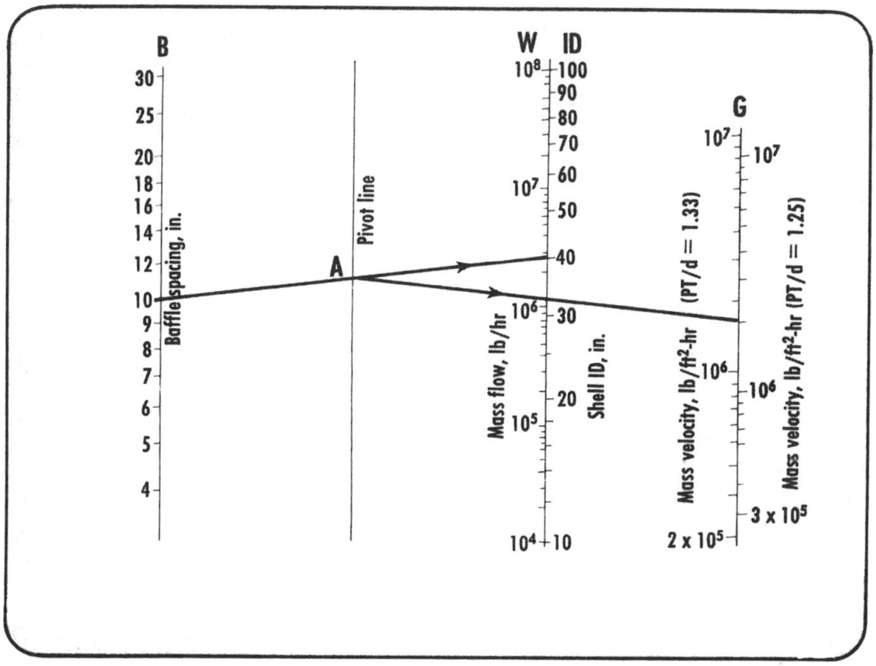


Fig. 1. Mass velocity nomogram

TWO NOMOGRAMS HAVE BEEN developed that offer a quick way to find shell-side heat-transfer coefficient for liquids.

The basic equation for shell-side coefficients for a triangular-pitch arrangement is:

$$Nu = 0.36 Re^{0.55} Pr^{0.33} \quad (1)$$

This equation may be rewritten to obtain the heat-transfer coefficient:

$$H = 1.1\,(G^{0.55}/d^{0.45}) \times (c^{0.33} K^{0.67}/\mu^{0.22}) \quad (2)$$

where:

$$G = W/as \quad (3)$$

$$as = ID \times B(1 - d/PT)/144 \quad (4)$$

Equation 2 is tedious to solve as it involves reference to thermal and transport proper-

**Nomenclature**

- Nu Nusselt's Number, $Hd/12K$
- Re Reynold's Number, $Gd/12\mu$
- Pr Prandtl's Number, $\mu c/K$
- $H$ Heat transfer coefficient, Btu/ft²-hr-°F
- $d$ Tube outer dia, in.
- $G$ Mass velocity, lb/ft²-hr
- ID Shell inner dia, in.
- $B$ Baffle spacing, in.
- $PT$ Pitch of tubes, in.
- $as$ Shell-side flow area, sq ft
- $W$ Mass flow of fluid, lb/hr
- $c$ Specific heat of liquid, Btu/lb-°F
- $\mu$ Viscosity, lb/ft-hr
- $K$ Thermal conductivity, Btu/ft-hr-°F

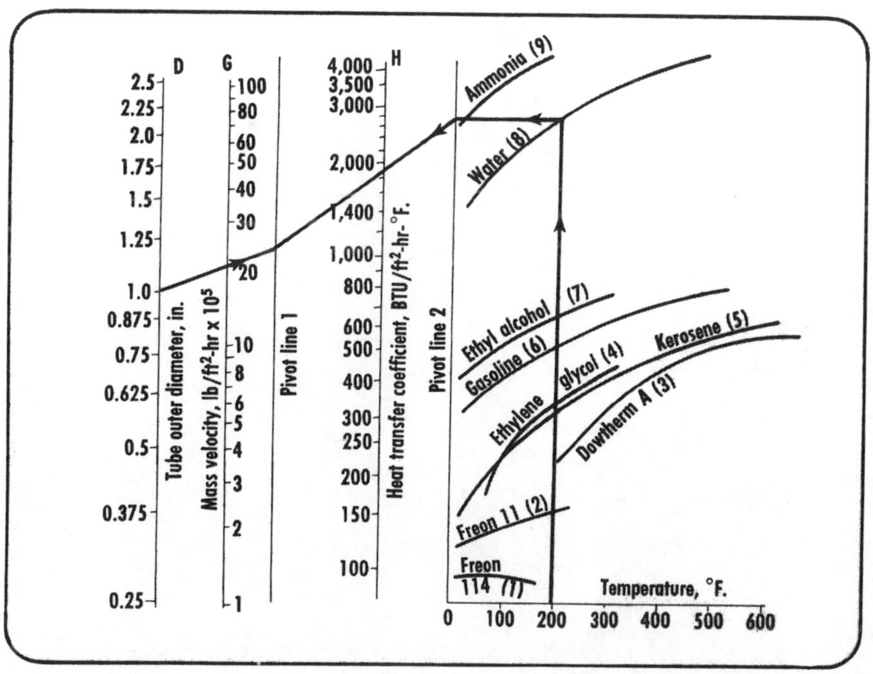


Fig. 2. Heat-transfer coefficients

ties of liquids. A simpler method is to calculate these and plot them on a chart as a function of temperature. This is done in Fig. 2, where the heat-transfer coefficient can be found if the temperature of the liquid, diameter of the tube, and mass velocity are known. Thus a lot of time is saved.

Fig. 1 is used to obtain a quick estimate of mass velocity, if mass flow, inner diameter of the heat exchanger, baffle pitch, and pitch-to-diameter ratio of the tubes are known. Two arrangements, $PT/d = 1.25$ and $1.33$, which are commonly used, are considered.

### *Example*

In a heat exchanger having a shell ID of 40 in., baffle pitch $B$ of 10 in., and a pitch-to-tube diameter ratio, $(PT/d)$ of 1.25, water flows at the rate 1.1 million lb/hr. What is the heat-transfer coefficient at 200°F, if the tube diameter is 1 in.?

### *Solution*

From Fig. 1, estimate mass velocity $G$. Connect $B = 10$ with ID = 40 to cut the pivot line at $A$. Connect with $W = 1.1 \times 10^6$ and extend to cut the $G$ scale.

Corresponding to $PT/d = 1.25$, read off $2 \times 10^6$ lb/ft$^2$-hr.

In Fig. 2, connect $d = 1.0$ with $G = 20 \times 10^5$ and extend to cut pivot line 1 at 1. Start from $T = 200$°F and go vertically to cut line 8 (water) and move horizontally to cut pivot line 2 at 2.

Connect with point 1, pivot line 1, to cut the $H$ scale. Read off 1,900 Btu/ft$^2$-hr-°F.

### *Bibliography*

Kern, D. Q., *Process Heat Transfer,* McGraw-Hill Book Co., New York.

# 4. Nomogram determines heat-transfer coefficient for water flowing in pipes or tubes

IN THE DESIGN OR EVALUATION OF HEAT exchangers, it is desirable to be able to make a quick estimate of heat transfer coefficients. This nomogram facilitates such estimates for film heat transfer coefficients for water flowing under forced turbulent conditions. It is based on the Dittus and Boelter's equations for heat flow from wall to liquid and from liquid to wall. These equations are given in terms of Reynolds number and Prandtl number. For quick use, they were rewritten in terms of directly measurable quantities, as follows:

$$h_h = (150+1.55T)V^{0.8}/D^{0.2}$$
$$h_c = (100+1.63T)V^{0.8}/D^{0.2}$$

where:

$T$ = temperature of water, F
$V$ = velocity of water, fps
$h_h$ = heat transfer coefficient for heating, Btu/hr-ft²-F
$h_c$ = heat transfer coefficient for cooling, Btu/hr-ft²-F

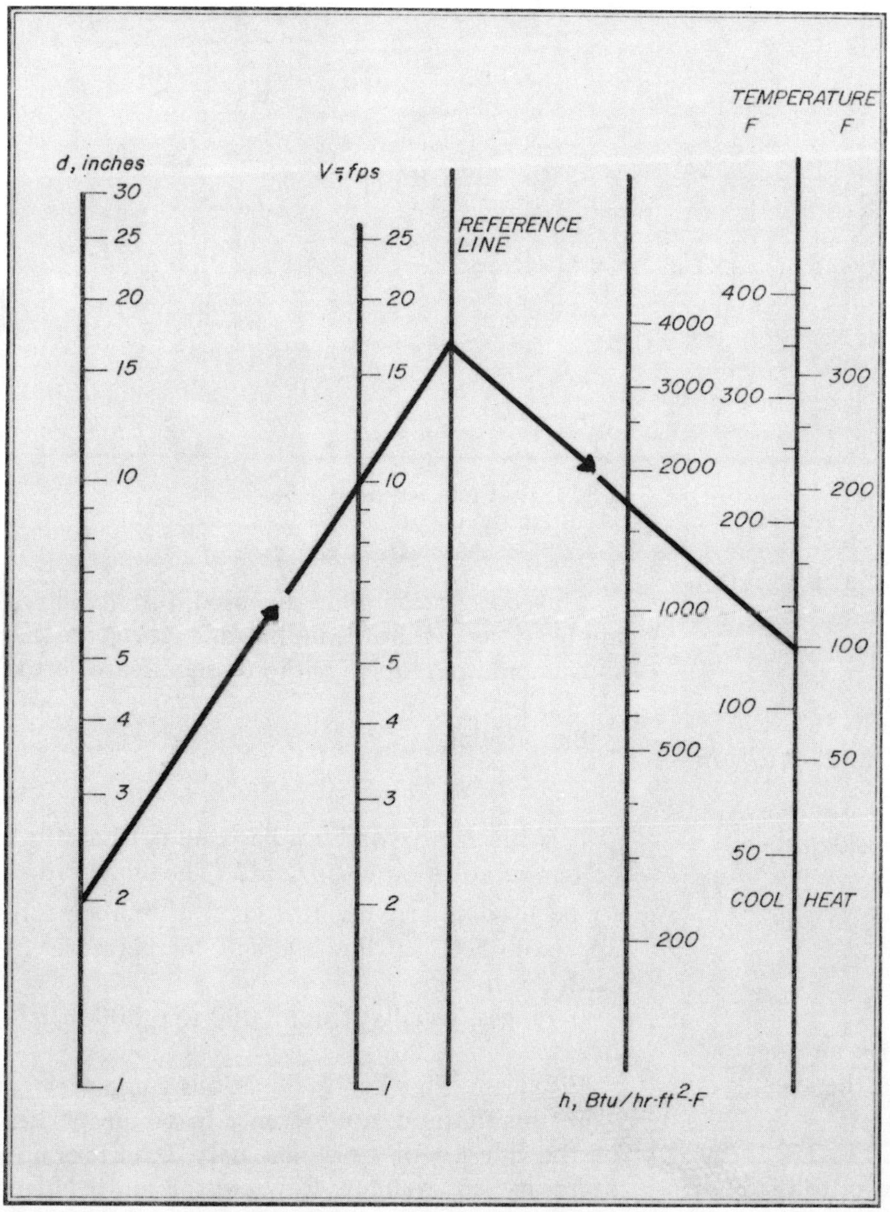


### *Example*

Water is heated in pipes of 2-in. inner diameter. Flow velocity is 10 fps and water temperature is 100F. Find the heat transfer coefficient.

### *Solution*

Connect $D = 2$ with $V = 10$ and extend to cut reference line. Connect this point with $T = 100$ and read answer on $h$-scale as 1700 Btu/hr-ft²-F.

Note that there are two temperature scales—one for heating and one for cooling.

# 5. Charts give quick estimates of heat-transfer coefficients for flow inside and outside tubes

AIR HEATERS AND WASTE heat recovery equipment frequently use hot flue gases to heat air.

Flue gases generally flow inside of the tubes while air flows over the tubes. Sometimes the flow pattern is reversed.

To determine the thermal performance it is desirable to be able to estimate the heat transfer coefficients inside and outside of the tubes.

Correlations widely used are presented below, using the following terms:

| | |
|---|---|
| $d$, $d_i$ | tube OD, ID, in. |
| $G$ | gas mass velocity, $lb/ft^2$-hr |
| $h_i$, $h_o$ | heat transfer coefficients inside and outside of tubes, $Btu/ft^2$-hr-°F |
| Nu | Nusselt number $= hd/12k$ |
| Pr | Prandtl number $= \mu cp/k$ |
| Re | Reynolds number $= 15.2\ W/(d_i\mu)$ or $Gd/(12\mu)$ |
| $W$ | flow inside each tube, lb/hr |
| $\mu$ | viscosity of fluid, lb/ft-hr |
| $Cp$ | specific heat of fluid, Btu/lb-°F |
| $k$ | thermal conductivity of fluid, Btu/ft-hr-°F |

For flow of air and flue gases inside tubes:

$$Nu = 0.023\ Re^{0.8}\ Pr^{0.4} \qquad (1)$$

Simplifying gives (Ref. 1):

$$h_i = 2.44\ W^{0.8}\ C/d_i^{1.8} \qquad (2)$$

$C$ is a function of $Cp$, $\mu$, and $k$ and depends on temperature. $C$ is evaluated at the average air or gas temperature.

$W$ lies usually in the range of 60 to 250 lb/hr.

For flow of air and flue gases outside tubes:

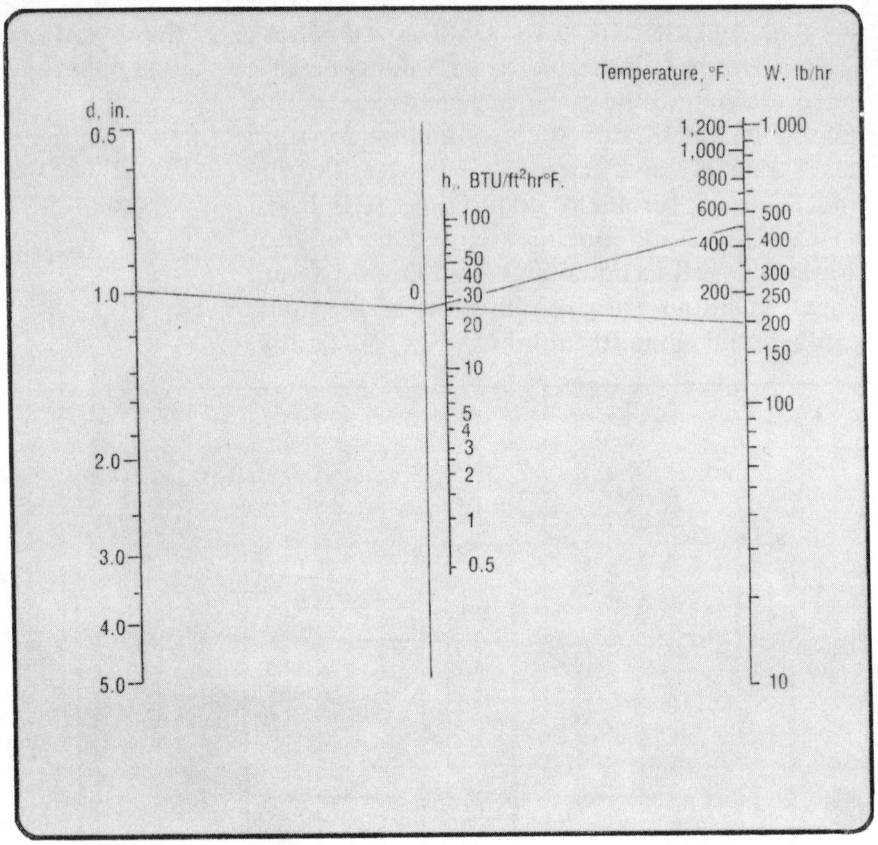


**Fig. 1. Heat transfer inside tubes**

$$Nu = 0.33\ Re^{0.6}\ Pr^{0.33} \qquad (3)$$

This correlation may be used for industrial applications of both in-line and staggered arrangement of tubes without significant error (Ref. 1).

Simplifying:

$$h_o = 0.9\ G^{0.6}\ F/d^{0.4} \qquad (4)$$

$F$ is the factor which takes care of the effect of temperature on $\mu$, $Cp$, and $k$. The temperature to be used in arriving at $h_o$ is film temperature approximated as the average temperature of both the fluids.

$G$ ranges, usually, from 3,000 to 6,000 $lb/ft^2$-hr.

Fig. 1 and Fig. 2 solve equations 2 and 4.

In this method, no reference to the properties of the flue gas or air is necessary. And it is not necessary to evaluate Reynolds, Prandtl Numbers.

***Example 1***

Estimate the heat transfer coefficient $h_i$ when 200 lb/hr of hot air at 500°F flows inside a tube with an ID of 1.0 in.

For the solution, connect $d$ = 1.0 with $W$ = 200 to cut reference line at 0. (Fig. 1)

Connect with temperature = 500 to cut $h_i$ scale at 29 Btu/ft²-hr-°F.

***Example 2***

Estimate $h_o$ when flue gas at a film temperature of 500 °F flows over a tube bundle of 1.5-in. OD tubes with a gas mass velocity of 6,000 lb/ft²-hr.

For the solution (Fig. 2), connect $G$ = 6,000 with $d$ = 1.5 to cut reference line at 0.

Now, connect with temperature = 500 to cut the $h_o$ scale at 15.2 Btu/ft²-hr-°F.

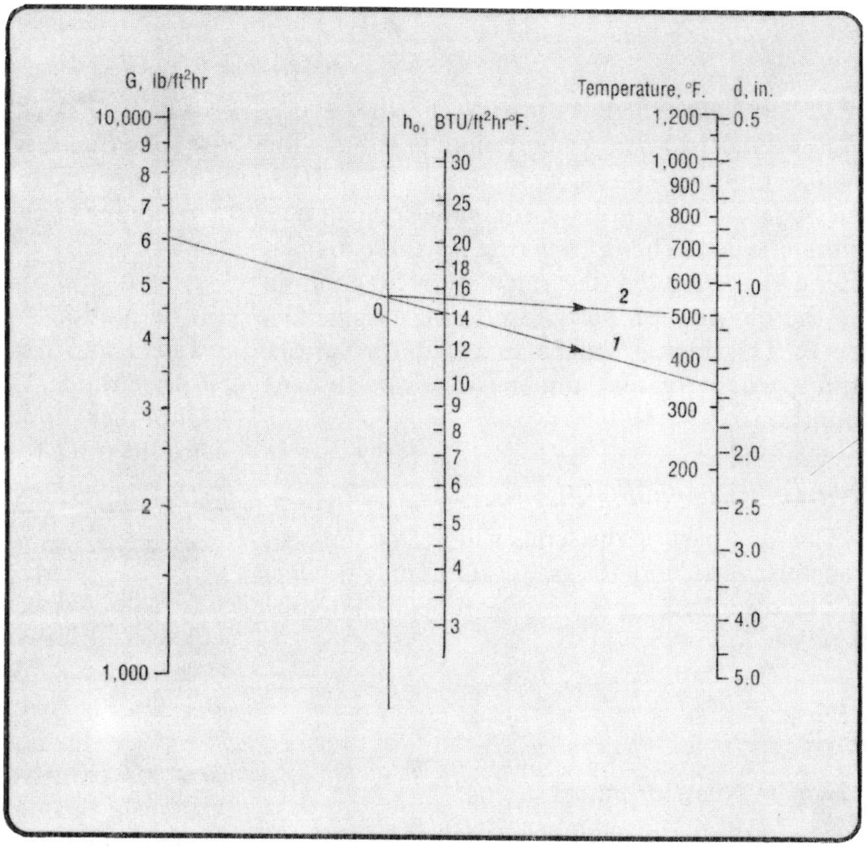


**Fig. 2. Heat transfer outside tubes**

***Reference***

1. V. Ganapathy, *Applied Heat Transfer,* PennWell Books, 1982.

# 6. Charts provide heat-transfer coefficients for hydrocarbons inside tubes

HEAT TRANSFER FOR A FLOWING FLUID USUALLY is estimated with an equation based on the fluid's physical properties raised to various fractional powers. Solving the equation can be a tedious affair. This is the first of three articles intended to simplify the calculation by combining the physical property terms into a single factor $C$. This factor will be reported for various liquids and gases under various flowing conditions.

### *Heat Transfer Coefficient*

This first part of the series will define the heat transfer coefficient for gases and liquids flowing in the turbulent region inside tubes. The equation is as follows:

$$N_{Nu} = 0.023\, N_{Re}{}^{0.8} N_{Pr}{}^{0.4} \quad (1)$$

where

$N_{Nu}$ = Nusselt number, $hD/k$
$N_{Re}$ = Reynolds number, $DG/\mu$
$N_{Pr}$ = Prandtl number, $C_p \mu/k$

Expanding Equation 1 to include the terms making up the dimensionless numbers, defining the tube diameter in inches and rearranging gives the following.

$$h = 2.44\,(W^{0.8}/d^{1.8})(k^{0.6} C_p{}^{0.4}/\mu^{0.4}) \quad (2)$$

where

$C_p$ = fluid specific heat, Btu/(lb) (°F)
$D$ = tube inside diameter, feet
$d$ = tube inside diameter, inches
$G$ = mass velocity, lb/(hr) (sq ft)
$h$ = heat transfer coefficient, Btu/(hr) (sq ft) (°F)
$k$ = thermal conductivity, Btu/(hr) (sq ft) (°F/ft)
$W$ = mass flow rate per tube, lb/hr = $G$ (area)
$\mu$ = fluid viscosity, lb/(ft) (hr) = 2.42 (centipoise)

Fluid properties shown in Equation 2 are combined into a single term as follows:

$$C = k^{0.6} C_p{}^{0.4}/\mu^{0.4} \quad (3)$$

where $C$ depends only on temperature $T$. Pressure effects will be considered from some gases in later parts of this series. Then Equation 2 becomes

$$h = 2.44\,(W^{0.8}/d^{1.8})(C) \quad (4)$$

Equation 4 is solved for some gases by using Fig. 1 and for some liquids by using Fig. 2. For other fluids, $C$ can be calculated by Equation 3 and then by entering the $C$ scale of either figure to solve for $h$.

### *Example 1*

What is the heat transfer coefficient for 1,000 lb/hr of ethylene glycol vapor flowing through a 0.75 in. ID tube at 200°F?

On Fig. 1, extend a line from $T = 200$ through a point having the coordinates $X = 3.5$ and $Y = 9.7$ to get $C = 0.19$. A second line from the $C$ scale through $W = 1{,}000$ gives a point on the reference line. A third line connecting the reference point and $d = 0.75$ will intersect at $h = 190$ Btu/(hr) (sq ft) (°F).

### *Example 2*

What is the heat transfer coefficient for 600 lb/hr of liquid propane flowing through a 1.25 in. ID tube at 32°F?

On Fig. 2, extend a line from $T = 32$ through a point having the coordinates $X = 8.5$ and $Y = 7.8$ to get $C = 0.26$. A second line from the $C$ scale through $W = 600$ to the reference line and then a third line from there to $d = 1.25$ gives $h = 72$ Btu/(hr) (sq ft) (°F).

### *Pressure*

The effect of pressure is not considered in these data, but Part 2 (section 7) will show how $C$ changes for 12 light gases at pressures up to 250 atmospheres (about 3,600 psia).

### *Bibliography*

Gallant, R. W., *Physical Properties of Hydrocarbons,* Vol. 1, Gulf Pub. Co., Houston, 1968.

**Table 1. Coordinates for Fig. 1, Gases**

| *Gas* | *X* | *Y* |
|---|---|---|
| Methane | 4.1 | 12.6 |
| Ethane | 3.7 | 11.6 |
| Propane | 3.4 | 11.6 |
| Butane | 3.4 | 11.0 |
| Ethylene | 3.2 | 10.8 |
| Propylene | 3.5 | 10.7 |
| Butene | 4.0 | 10.3 |
| Methanol | 3.9 | 9.2 |
| Ethanol | 3.2 | 10.3 |
| Propanol | 3.2 | 10.3 |
| Butanol | 3.3 | 10.1 |
| Ethylene glycol | 3.5 | 9.7 |
| Diethylene glycol | 3.5 | 9.3 |
| Triethylene glycol | 3.8 | 9.0 |
| Chloroform | 5.4 | 2.7 |
| Methylchloride | 4.2 | 4.9 |
| Carbon tetrachloride | 6.1 | 0.7 |

**Table 2. Coordinates for Fig. 2, Liquids**

| *Liquid* | *X* | *Y* |
|---|---|---|
| Methane | 5.8 | 10.9 |
| Ethane | 8.9 | 8.6 |
| Propane | 8.5 | 7.8 |
| Butane | 9.0 | 7.3 |
| Isopentane | 8.3 | 6.5 |
| Ethylene | 9.8 | 8.5 |
| Propylene | 9.8 | 8.3 |
| Butene-1 | 9.3 | 7.4 |
| Isoprene | 8.8 | 7.2 |
| Methanol | 7.0 | 6.4 |
| Ethanol | 6.0 | 5.5 |
| Propanol | 5.0 | 4.8 |
| Butanol | 4.3 | 4.7 |
| Methylene chloride | 8.0 | 5.7 |
| Carbon tetrachloride | 7.1 | 4.3 |

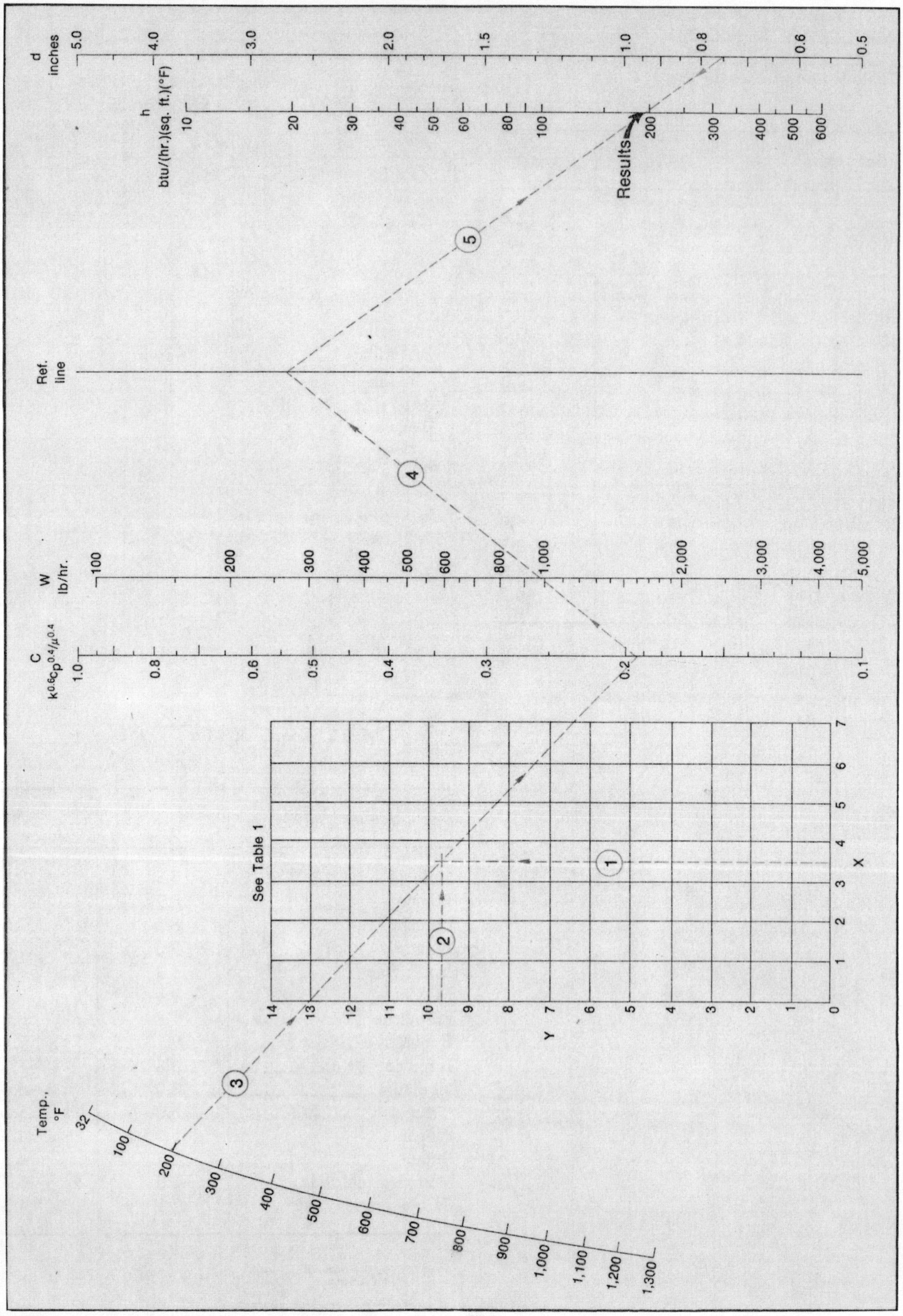


**Fig. 1. For gases with coordinates given in Table 1 to solve for heat-transfer coefficient**

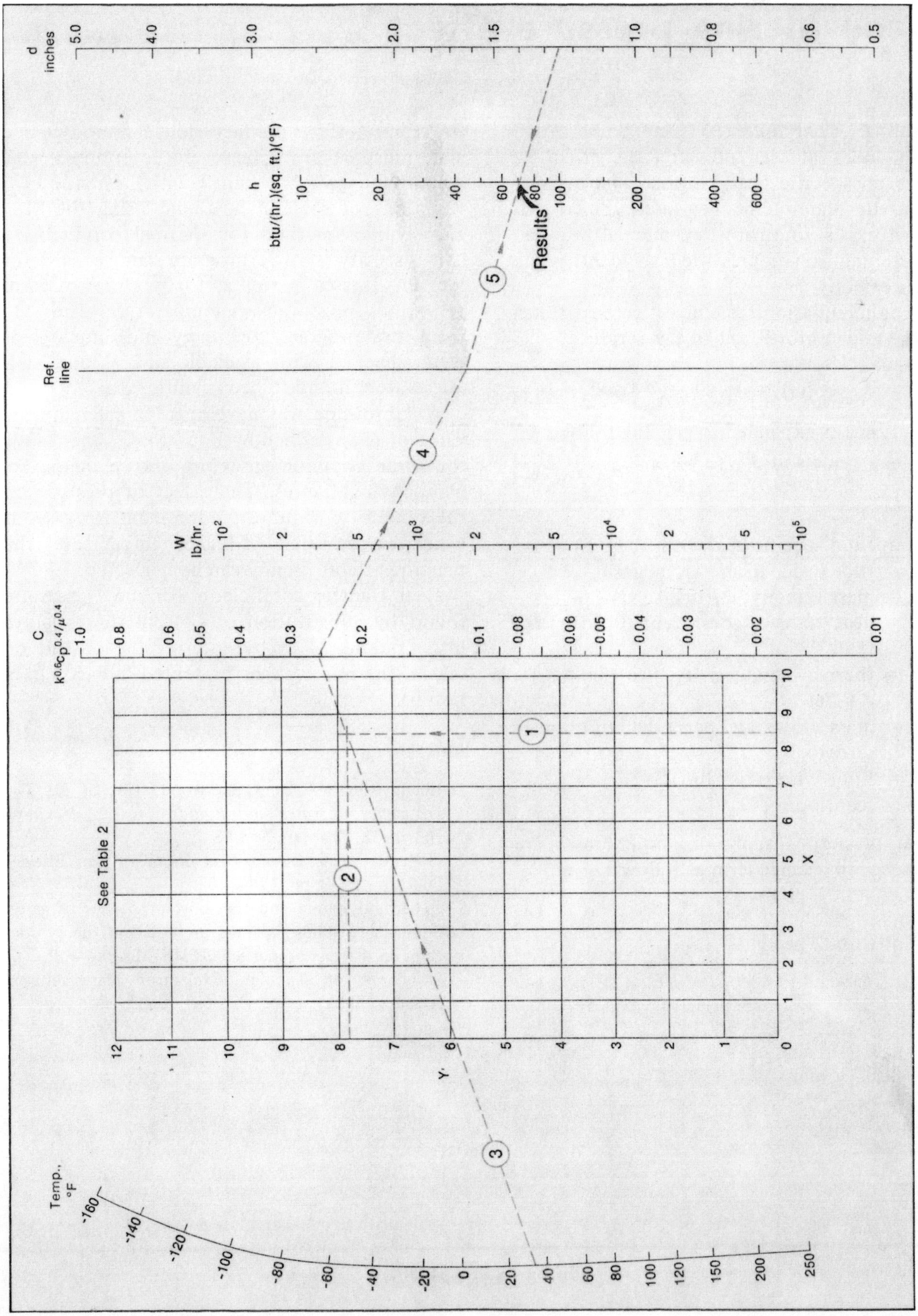


**Fig. 2. For liquids with coordinates given in Table 2 to solve for heat-transfer coefficient**

# 7. Charts give heat-transfer coefficients considering pressure effect

FOR HEAT TRANSFER BY GASES FLOWING through tubes at high pressure, the effect of pressure on specific heat, viscosity and thermal conductivity cannot be neglected. Since this situation exists for many commercial heat exchangers, graphs are presented to identify the pressure effect.

The basic equation for convective heat transfer is the same as for Part 1 of this series:

$$N_{Nu} = 0.023\, N_{Re}^{0.8} N_{Pr}^{0.4} \quad (1)$$

Equation 1 is expanded to give the following:

$$h = 2.44\,(W^{0.8}/d^{1.8})(k^{0.6} C_p^{0.4}/\mu^{0.4}) \quad (2)$$

where

$C_p$ = fluid specific heat, Btu/(lb) (°F)
$d$ = tube inside diameter, inches
$G$ = mass velocity, lb/(hr) (sq ft)
$h$ = heat transfer coefficient, Btu/(hr) (sq ft) (°F)
$k$ = thermal conductivity, Btu/(hr) (sq ft) (°F/ft)
$W$ = mass flow rate per tube, lb/hr = $G$ (area)
$\mu$ = fluid viscosity, lb/(ft) (hr) = 2.42 (centipoise)

Fluid properties shown in Equation 2 are combined into a single term as follows:

$$C = k^{0.6} C_p^{0.4}/\mu^{0.4} \quad (3)$$

Then Equation 2 becomes

$$h = 2.44\,(W^{0.8}/d^{1.8})(C) \quad (4)$$

Pressure effects on the various thermodynamic and transport properties were determined with a computer program using Gambill's method for variations of specific heat and using the Stiel and Thodos method for thermal conductivity and viscosity.[1,2] The program can be used for individual gases as well as for mixtures of gases at various pressure and temperature. The data from the program for many individual gases were checked with available data[3,4] and variations were found to be insignificant.

A subroutine was developed to calculate the value of $C$ as defined by Equation 3. The results for some common gases are shown in the accompanying figures. The effect of pressure on the value of $C$ is more predominant for gases at lower temperatures. At higher temperatures, the pressure effects usually can be neglected.

Heat transfer coefficients for these gases are found by substitution of $C$ from these figures into Equation 4 or by entering the $C$ scale on one of the nomographs presented in Part 1 (section 6).

### References


1. Reed, Robert C. and Sherwood, Thomas K., *The Properties of gases and liquids,* 1st Ed., McGraw-Hill, 1958, Chapters 7 and 8.
2. Holland, F. A., et al., *Heat transfer,* Am. Elsevier Pub. Co., 1970, pp. 324–340.
3. "Thermodynamic and transport properties of gases, liquids and solids," papers presented at the symposium on thermal properties, ASME, McGraw-Hill.
4. Standards of Tubular Exchanger Manufacturers Assoc. (TEMA).

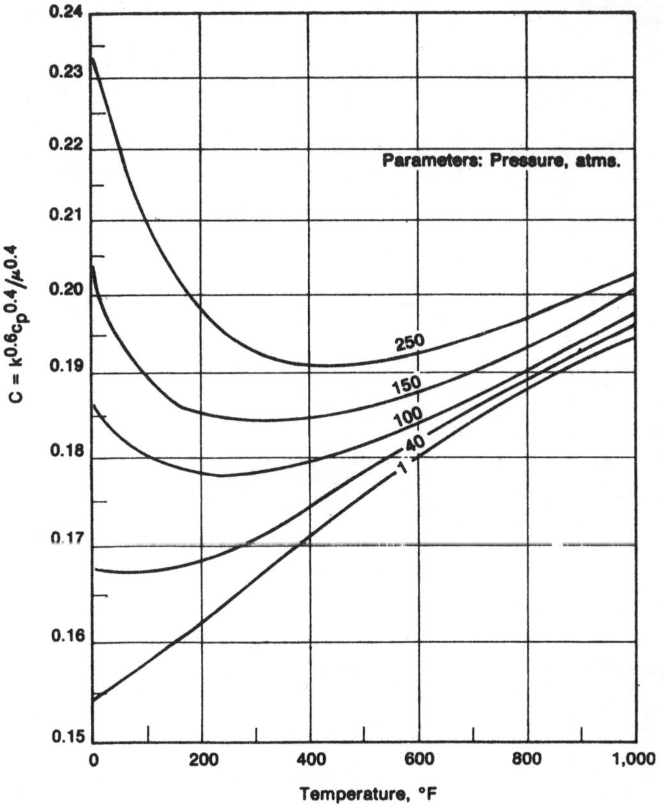


Fig. 1. Air

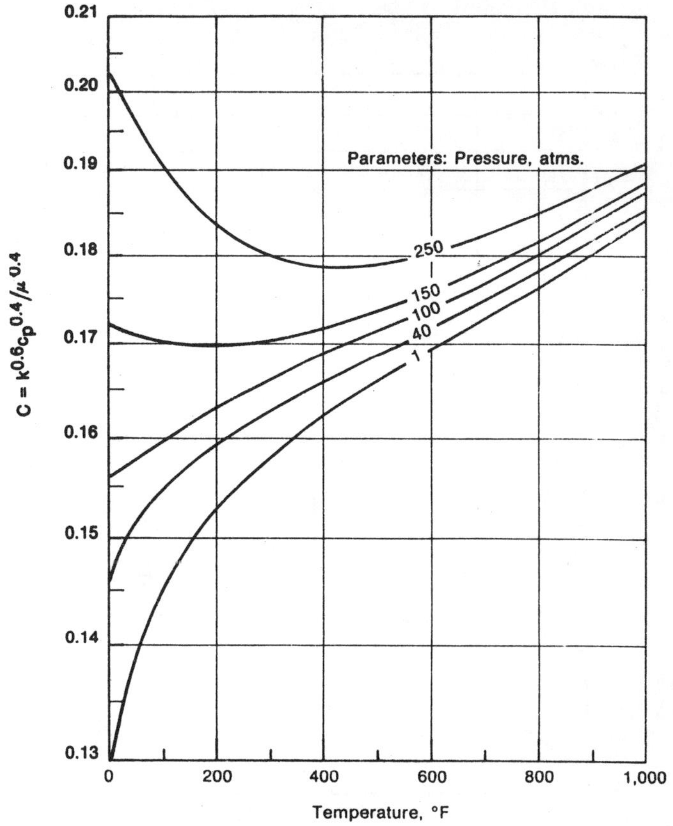


Fig. 2. Oxygen

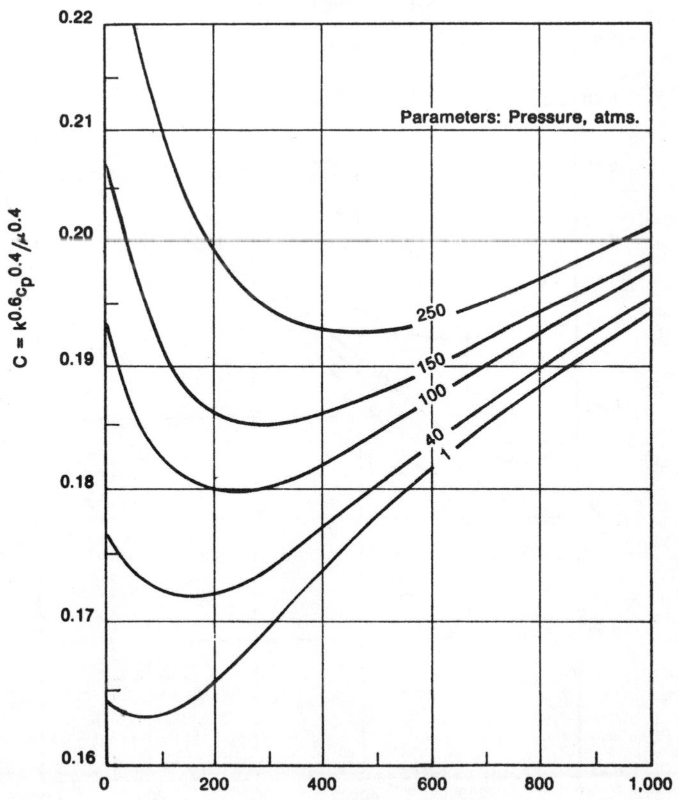


Fig. 3. Nitrogen

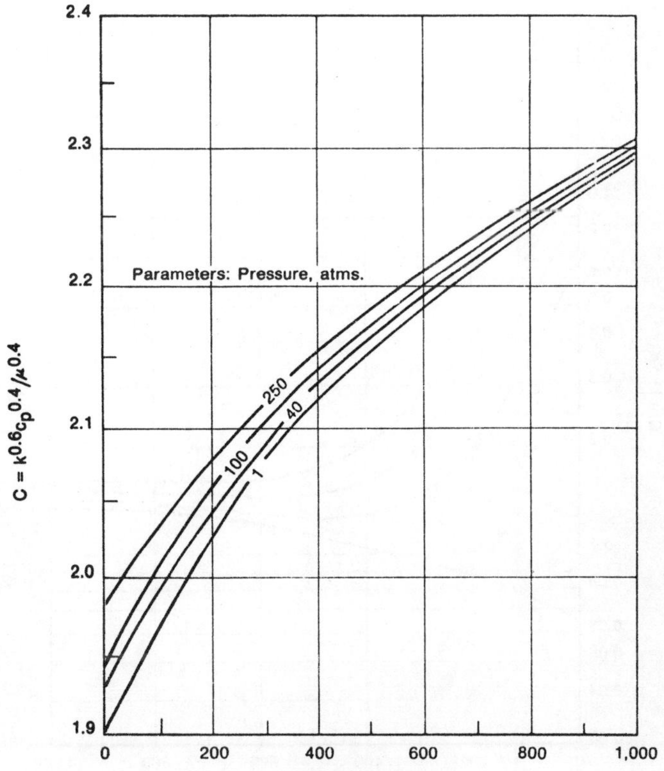


Fig. 4. Hydrogen

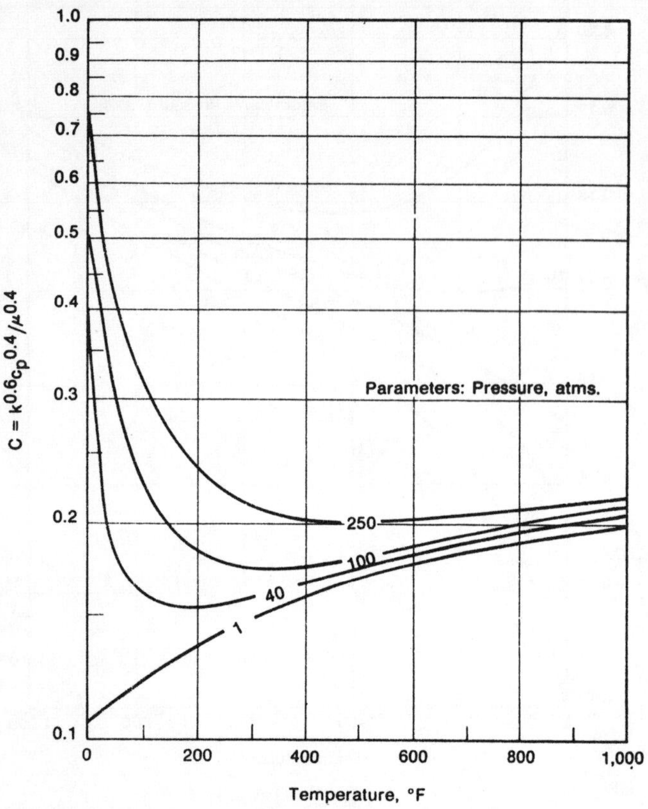


**Fig. 5. Carbon dioxide**

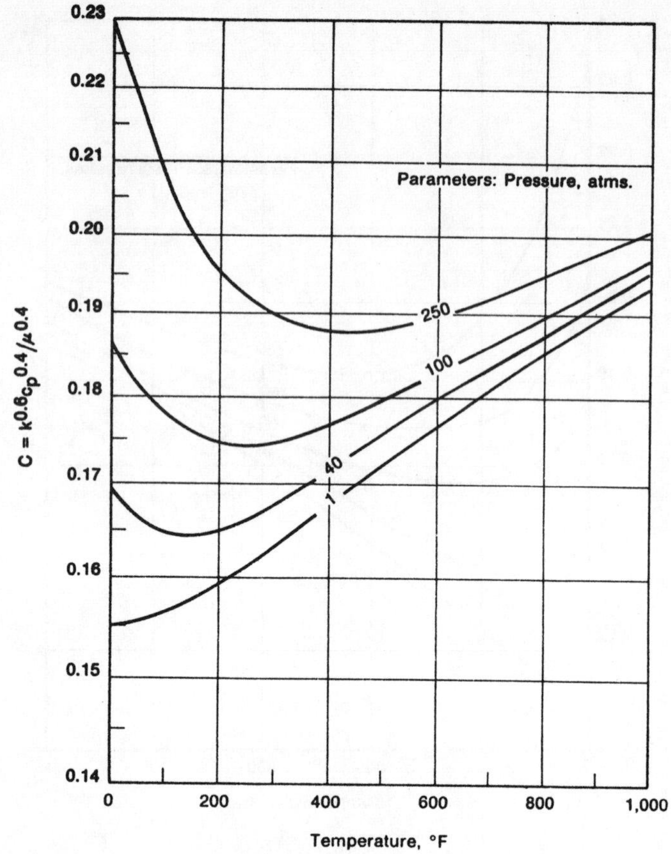


**Fig. 6. Carbon monoxide**

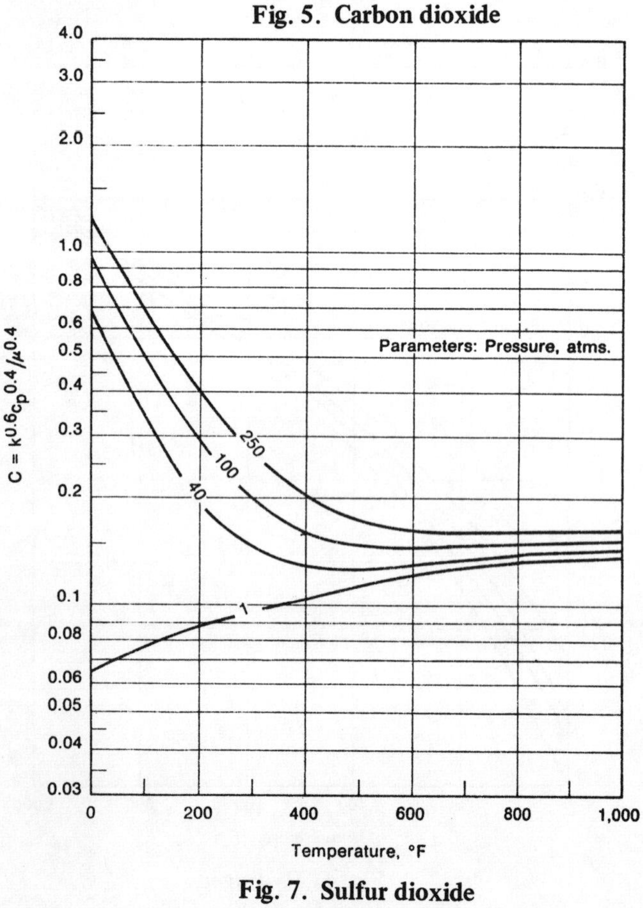


**Fig. 7. Sulfur dioxide**

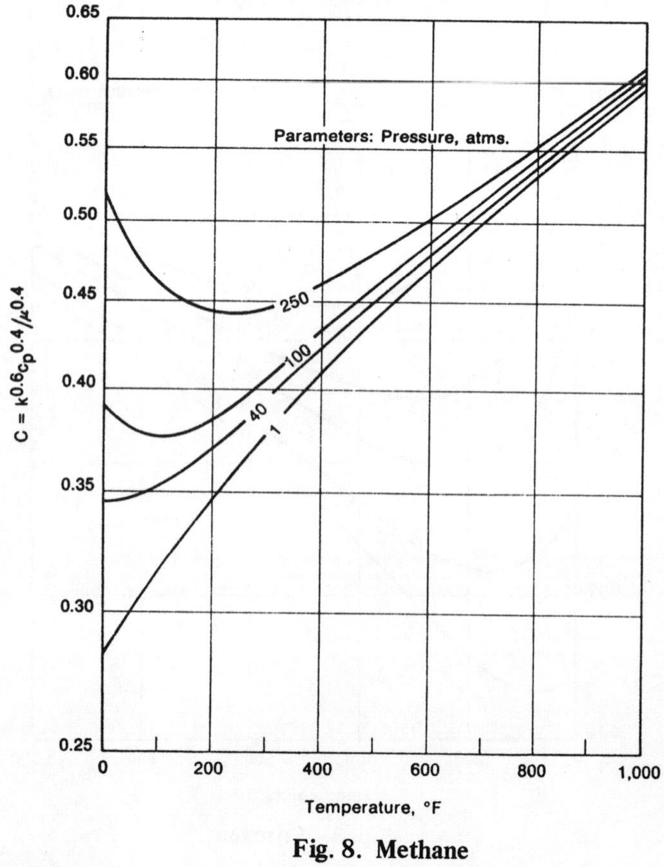


**Fig. 8. Methane**

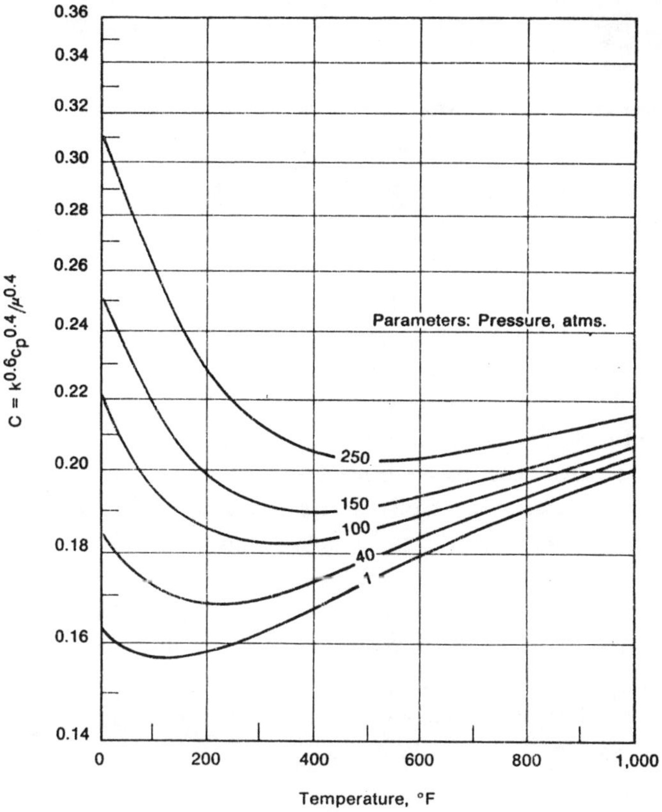


Fig. 9. Nitric oxide

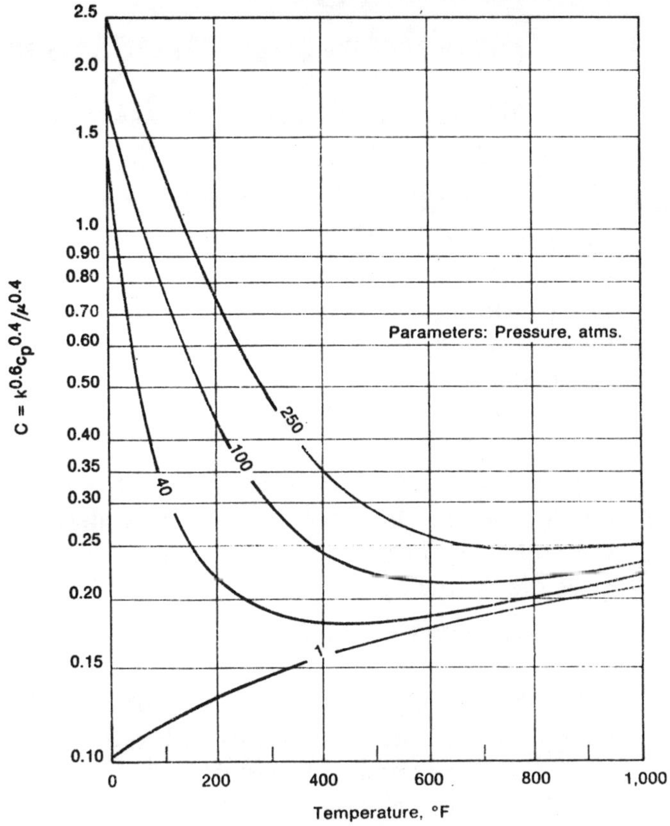


Fig. 10. Nitrogen peroxide

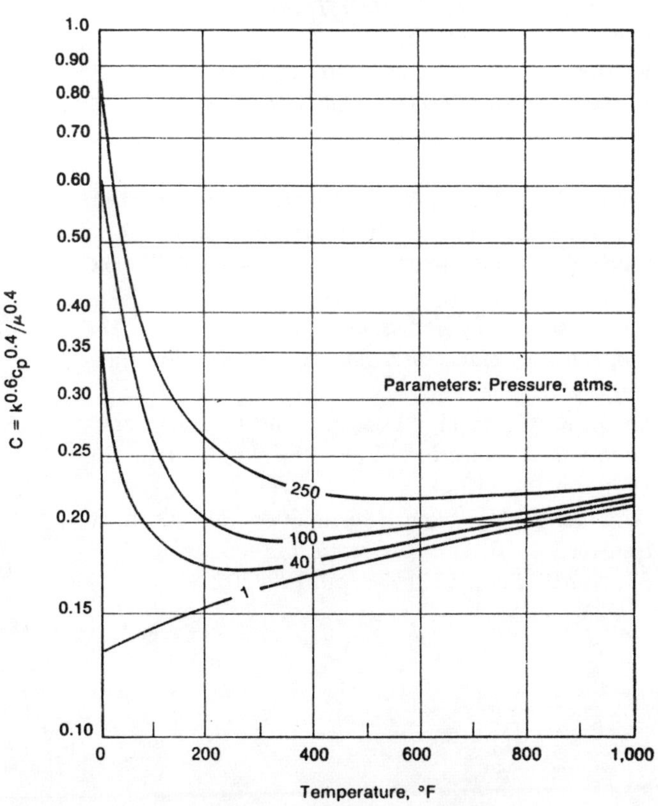


Fig. 11. Nitrous oxide

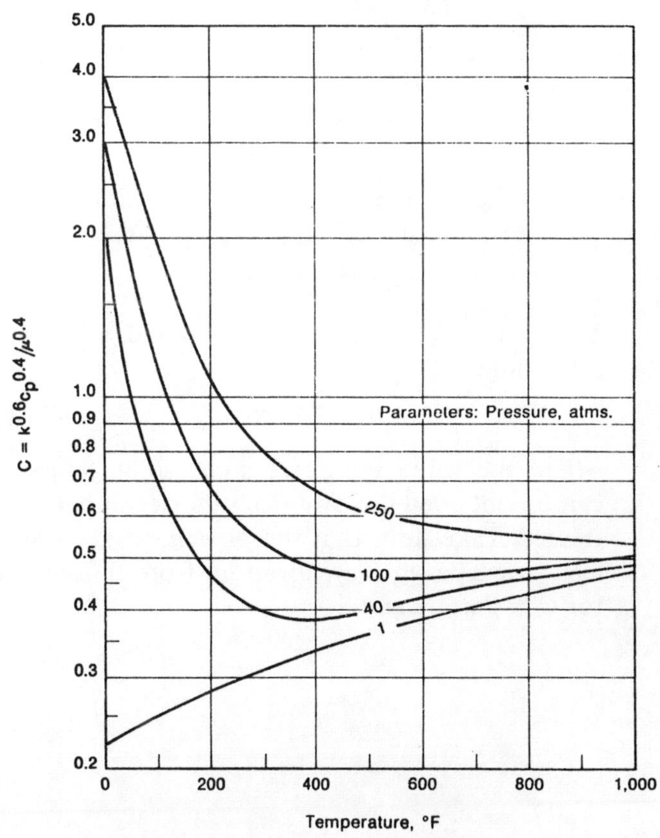


Fig. 12. Ammonia

# 8. Charts give heat-transfer coefficients considering pressure effect in crossflow situations

HEAT TRANSFER COEFFICIENTS FOR FLUIDS flowing over tube bundles are calculated with a slightly different equation than the one used in Parts 1 and 2. For a staggered arrangement of tubes, the equation is as follows:

$$N_{Nu} = 0.33\, N_{Re}{}^{0.6} N_{Pr}{}^{0.33} \qquad (1)$$

Expanding Equation 1 to include the terms making up the dimensionless numbers, defining the tube diameter in inches and rearranging gives the following:

$$h = 0.9\,(G^{0.6}/d^{0.4})(k^{0.67}\,C_p{}^{0.33}/\mu^{0.27}) \qquad (2)$$

where

- $C_p$ = fluid specific heat, Btu/(lb)(°F)
- $d$ = tube outside diameter, inches
- $G$ = mass velocity, lb/(hr)(sq ft)
- $h$ = heat transfer coeff., Btu/(hr)(sq ft)(°F)
- $k$ = thermal conductivity, Btu/(hr)(sq ft)(°F/ft)
- $\mu$ = fluid viscosity, lb/(ft)(hr)=2.42 (centipoise)

Fluid properties shown in Equation 2 are combined into a single term as follows:

$$F = k^{0.67}\,C_p{}^{0.33}/\mu^{0.27} \qquad (3)$$

Then the heat transfer is calculated using the following:

$$h = 0.9\,(G^{0.6}/d^{0.4})(F) \qquad (4)$$

where the value for $F$ for some common gases can be obtained directly from the accompanying figures. Take note that the powers for the physical properties in $F$ are different from those used for $C$ in Parts 1 and 2.

The figures for $F$ are the results of a computer program in which pressure effect on specific heat is determined using Gambill's method and the effects on thermal conductivity and viscosity are by the method of Stiel and Thodos.

The figures cover a wide range of pressures; namely, 1 to 250 atmospheres. Hence they can be used for the more common cases at atmospheric pressure as well as for the higher pressures encountered in waste heat recovery applications.

For in-line arrangement of tubes, the value obtained for heat-transfer coefficient by the equations herein should be multiplied by 0.8. For a baffled heat exchanger, an additional multiplier of 0.6 should be used.

### *Bibliography*


1. McAdams, William H., *Heat Transmission,* 3rd Ed., McGraw-Hill, 1954.
2. Rohsenow, Warren and Hartnett, J. P., *Handbook of Heat Transfer,* McGraw-Hill, 1972.
3. Kern, Donald Q., *Process Heat Transfer,* McGraw-Hill, 1972.
4. Perry, Robert H. and Chilton, Cecil H., *Chemical Engineers' Handbook,* 5th Ed., McGraw-Hill, 1973.
5. "Steam, its generation and use," Babcock & Wilcox, USA.
6. Reed, Robert C. and Sherwood, Thomas K., *The Properties of Gases and Liquids,* 2nd Ed., McGraw-Hill, 1966.
7. Lord, R. C., et al, "Design of heat exchangers," *Chemical Engineering,* Vol. 77, No. 2, Jan. 26, 1970, pp. 96–118.
8. Ganapathy, V., "Quick estimation of gas heat-transfer coefficients," *Chemical Engineering,* Vol. 83, No. 19, Sept. 13, 1976, pp. 199–202.

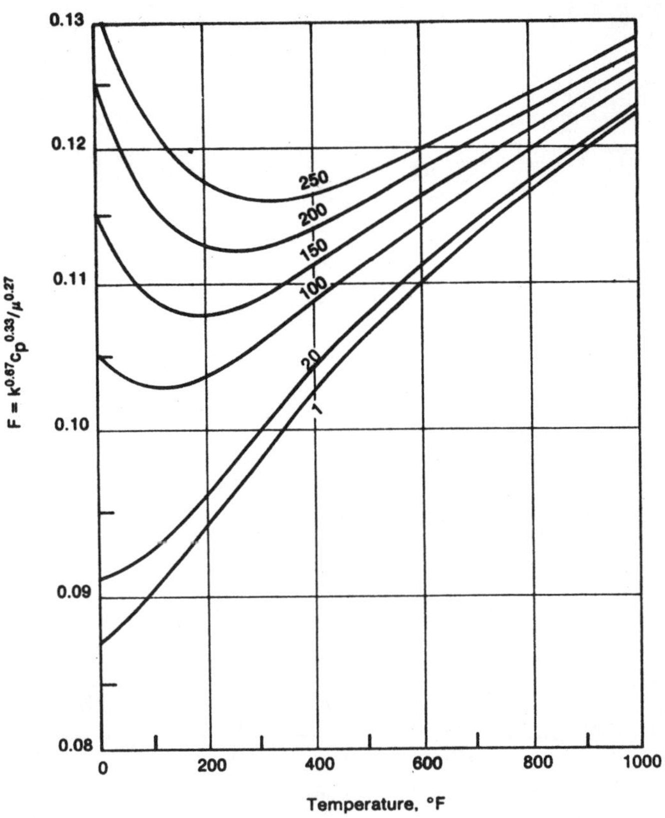


Fig. 1. Air

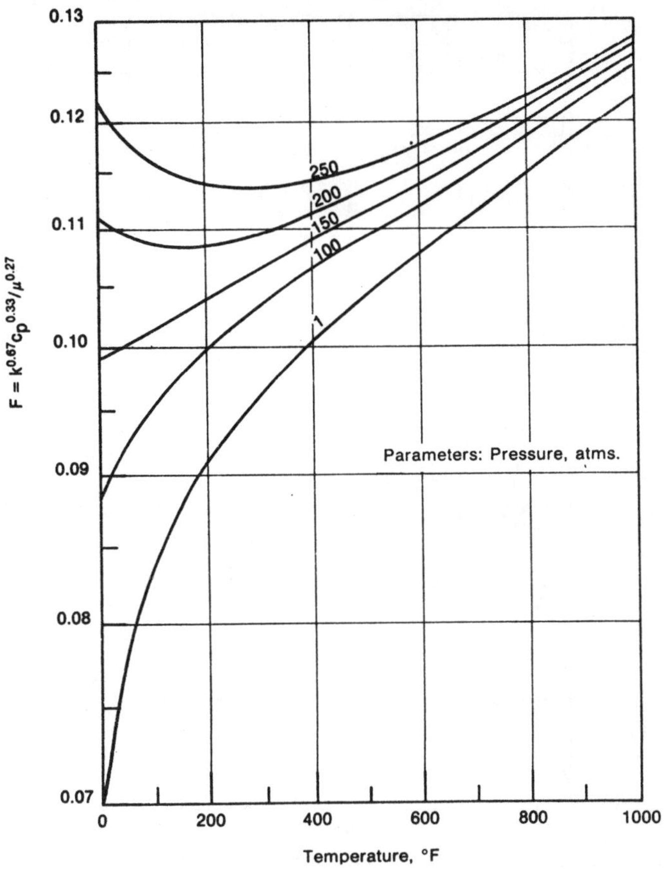


Fig. 2. Oxygen

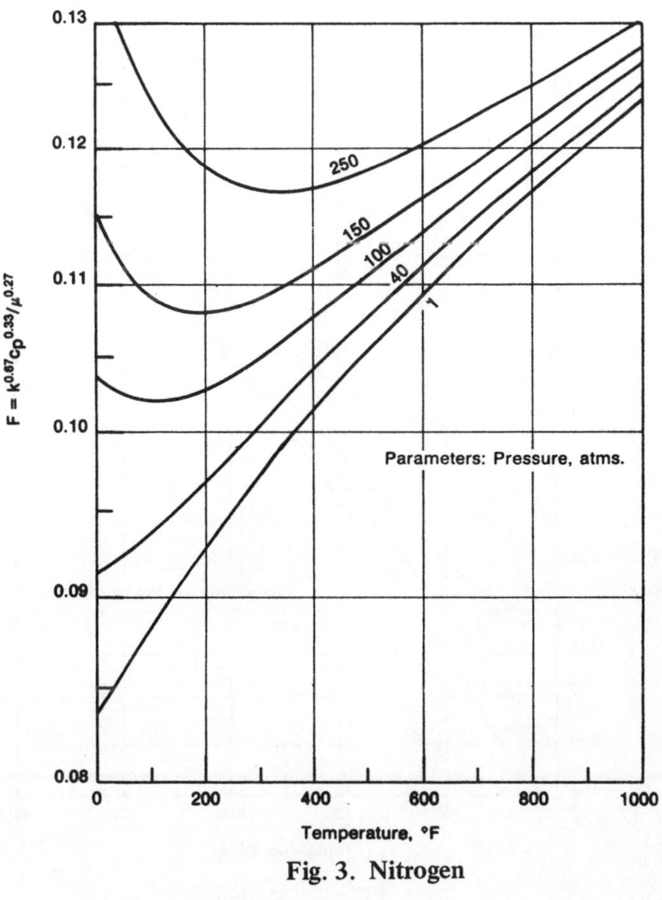


Fig. 3. Nitrogen

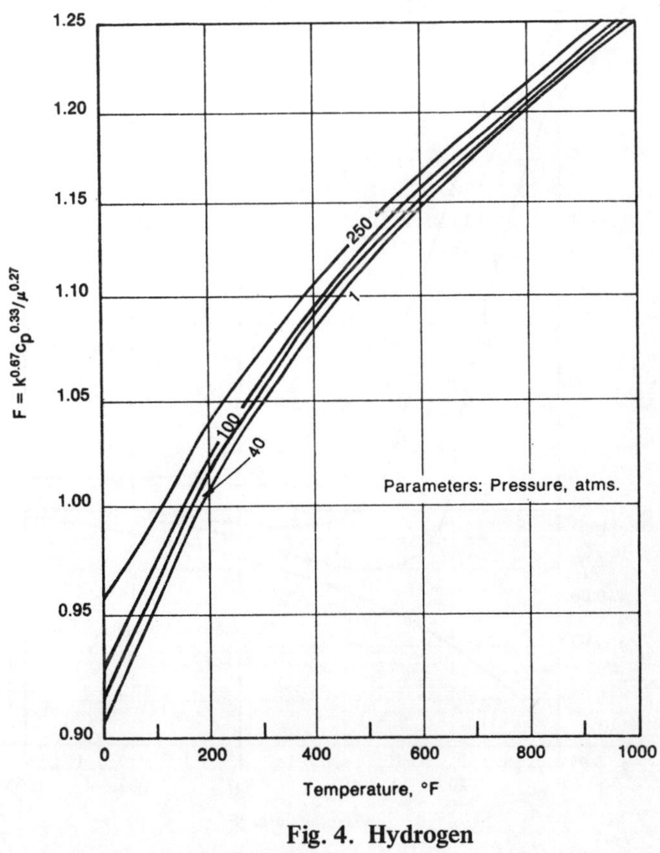


Fig. 4. Hydrogen

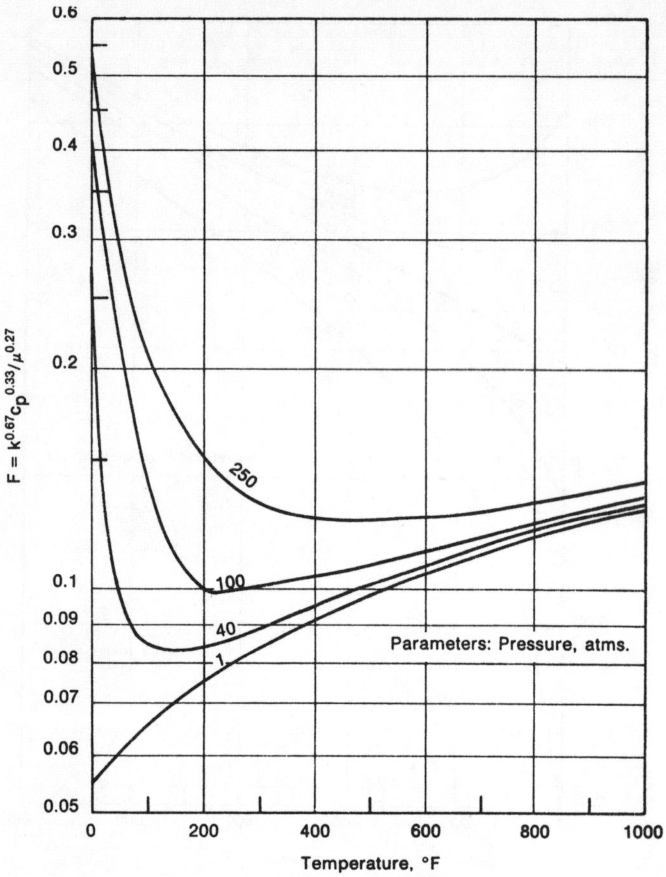


Fig. 5. Carbon dioxide

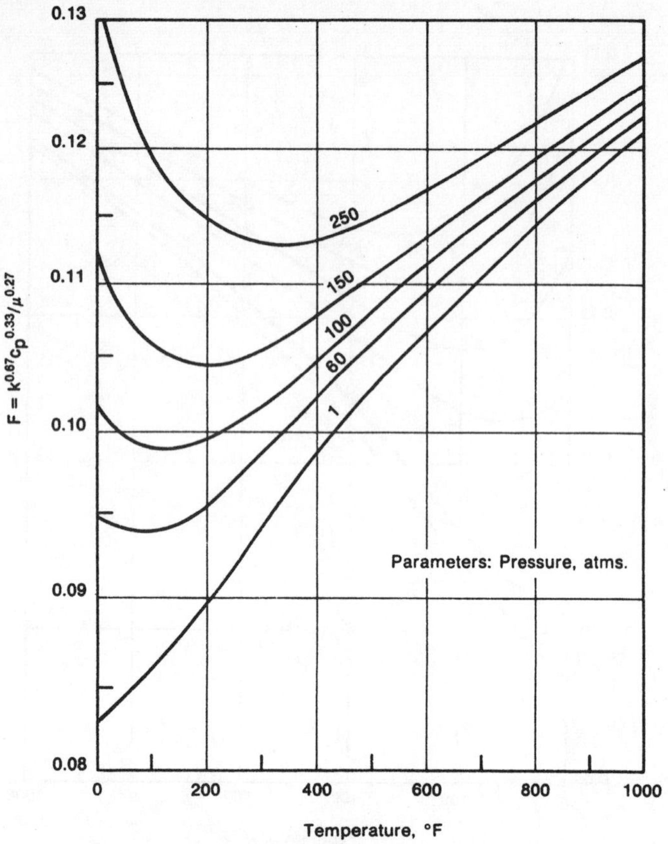


Fig. 6. Carbon monoxide

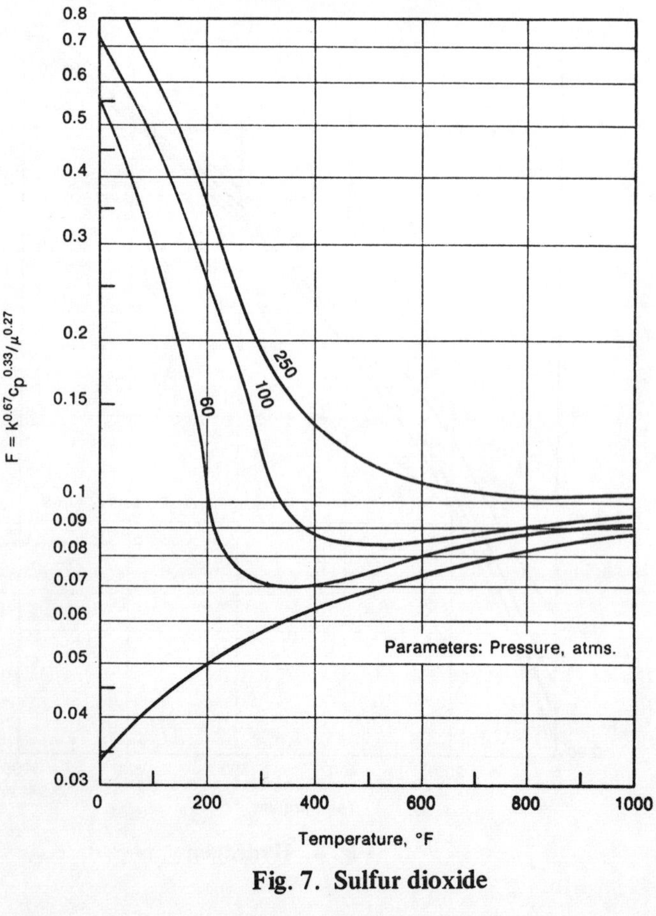


Fig. 7. Sulfur dioxide

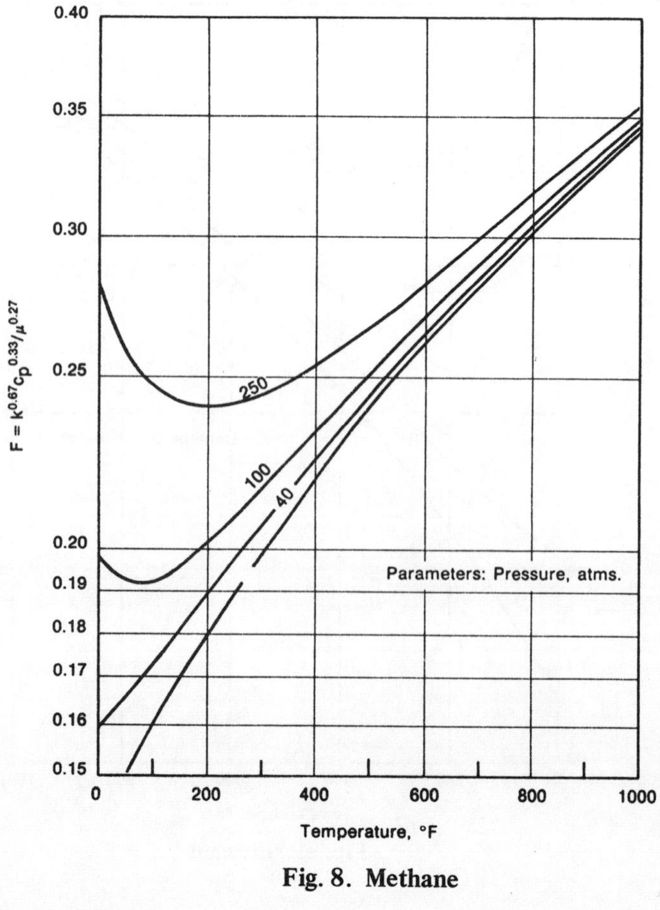


Fig. 8. Methane

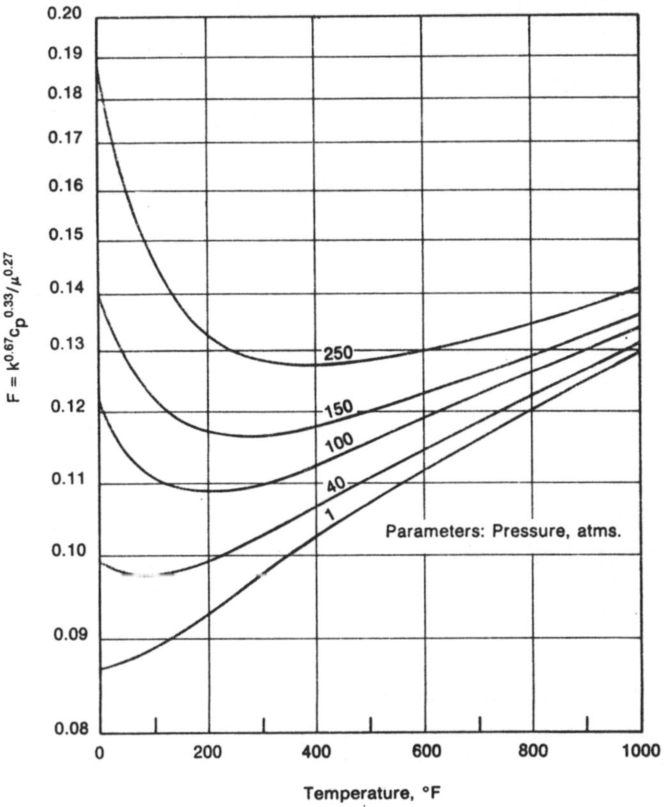


Fig. 9. Nitric oxide

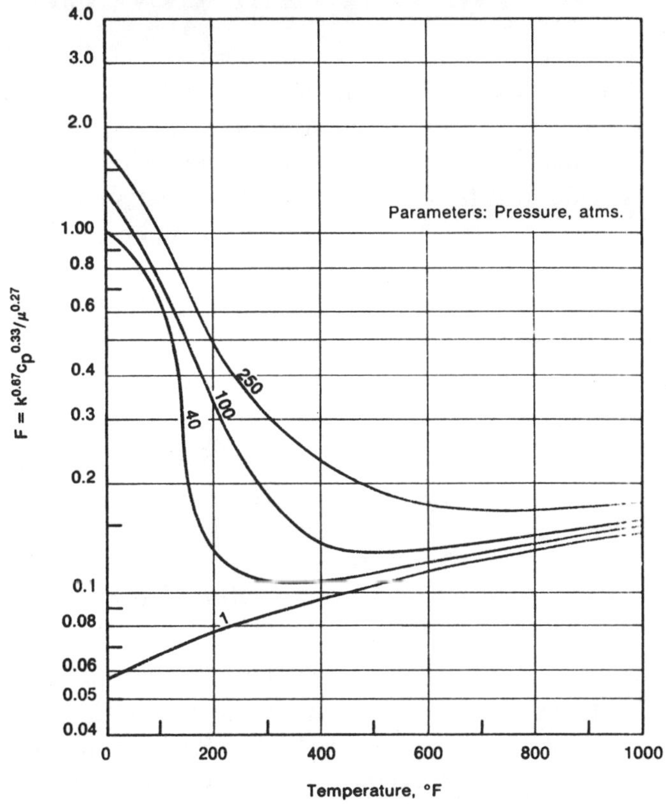


Fig. 10. Nitrogen peroxide

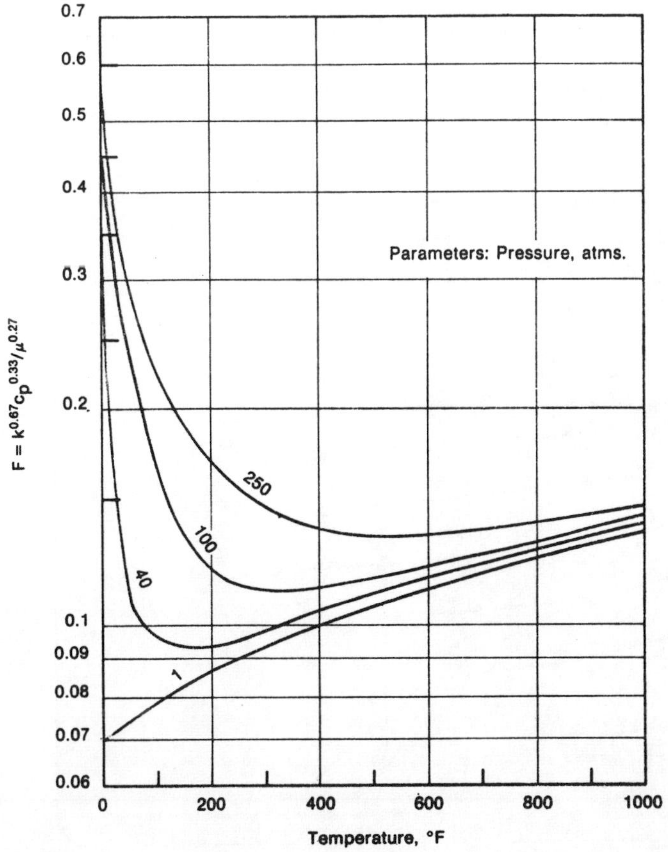


Fig. 11. Nitrous oxide

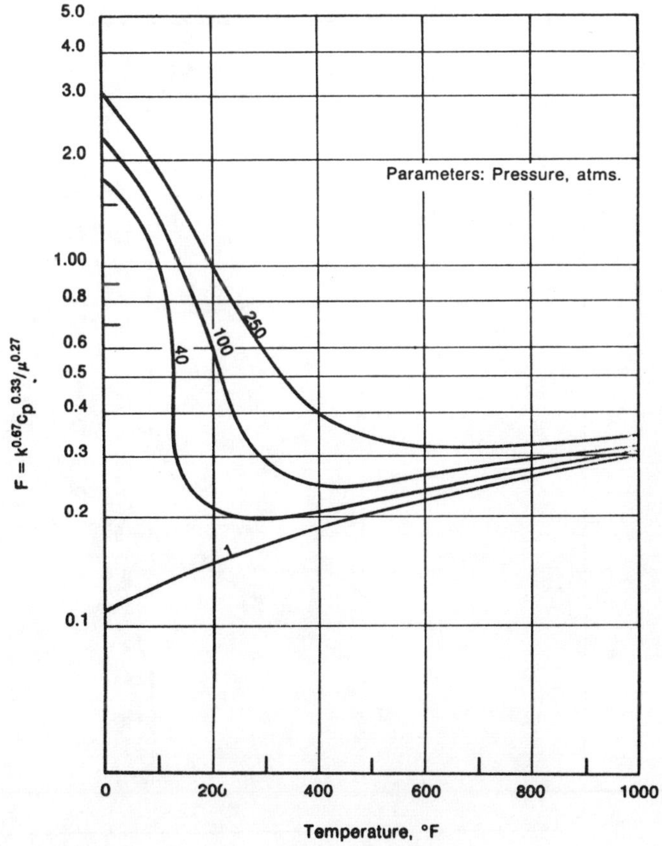


Fig. 12. Ammonia

# 9. Nomogram determines tube thermal resistance

RESISTANCE TO HEAT TRANSFER OF TUBE materials cannot be neglected in heat-transfer calculations, especially when the inside and outside heat-transfer coefficients are large, as is the case in liquid-to-liquid heat exchangers.

Metal thermal resistance due to tube wall thickness can be calculated using the following expression:

$$R = (d/24k)[\ln(D/d)]$$

where:

$R$ = thermal resistance due to wall thickness, ft²-hr-°F/Btu
$D$ = OD of tube, in.
$d$ = ID of tube, in.
$k$ = thermal conductivity of tube, Btu/ft-hr-°F

The preceding expression for thermal resistance is tedious to calculate when tube wall thickness is given in BWG (Birmingham Wire Gauge). The accompanying nomogram can be used to quickly obtain thermal resistance $R$, if $D$, $d$, and $k$ are known.

With the nomogram, there is no need to calculate the inner diameter. It is sufficient if the outer diameter and wall thickness in BWG are known. The thermal conductivity of several common tube materials is indicated for a temperature of 70°F.

### *Example*

Estimate the thermal resistance to heat transfer when 1¼-in. OD stainless steel tubes of 16 BWG are used in a heat exchanger operating at about 70°F.

### *Solution*

Start from $D$=1.25, go vertically to cut the 16 BWG curve. Move horizontally to cut Pivot line 1.

Connect that point with point $B$ on the $k$ scale and extend to cut the $R$ scale.

Read thermal resistance, $R$, as $5.8 \times 10^{-4}$ ft²-hr-°F/Btu.

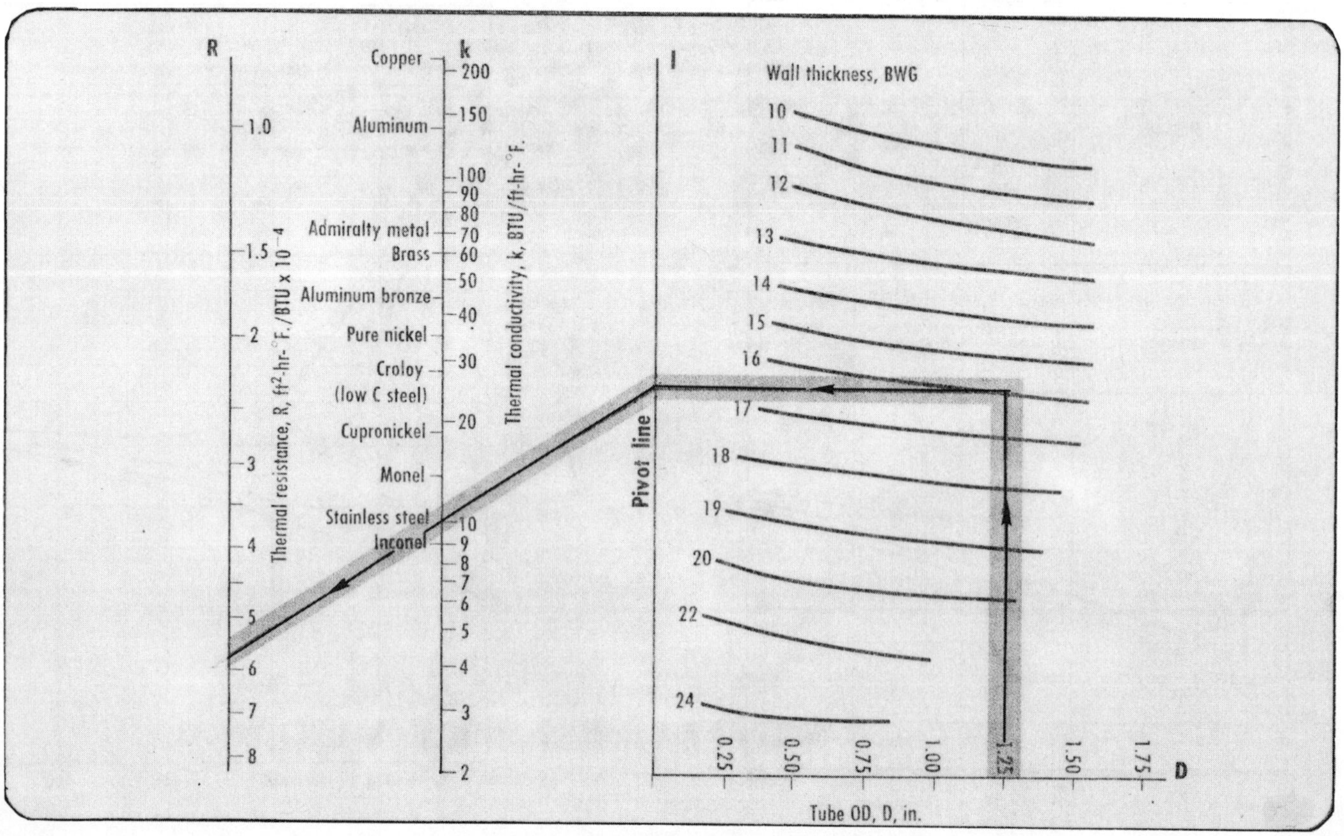

# 10. Chart provides quick estimate of boiling-heat-transfer coefficient

THIS NOMOGRAPH ESTIMATES THE HEAT-TRANSFER coefficient *(h)* for subcooled boiling of water inside tubes. The nomograph is based on an equation that predicts water-side film-temperature drop ($\Delta T$), given the heat flux across the tubes ($\phi$) and the pressure of the steam-water mixture *(P)*:

$$\Delta T = 0.072 e^{-P/1{,}260} \sqrt{\phi}$$

Once $\Delta T$ is computed, the heat-transfer coefficient *(h)* and the tubewall temperature ($T_w$) can be calculated by means of the following equations:

$$h = \phi / \Delta T$$

$$T_w = T_{sat} + \Delta T$$

### *Example*

What is the wall temperature of tubes in a boiler if heat flux is 60,000 Btu/ft²h and the pressure of the steam-water mixture is 1,200 psia? What is the boiling-heat-transfer coefficient?

### *Solution*

On the nomograph, connect $\phi$=60,000 with $P$=1,200 and extend the line to cut the $h$ scale. Read $h$=9,000 Btu/ft²h°F on the $h$ scale, and $\Delta T$=7°F on the $\Delta T$ scale. The wall temperature $T_w$=567+7=574°F.

### *Reference*


Collier, J. G., *Convective Boiling and Condensation*, McGraw-Hill, New York, 1972, p. 160.


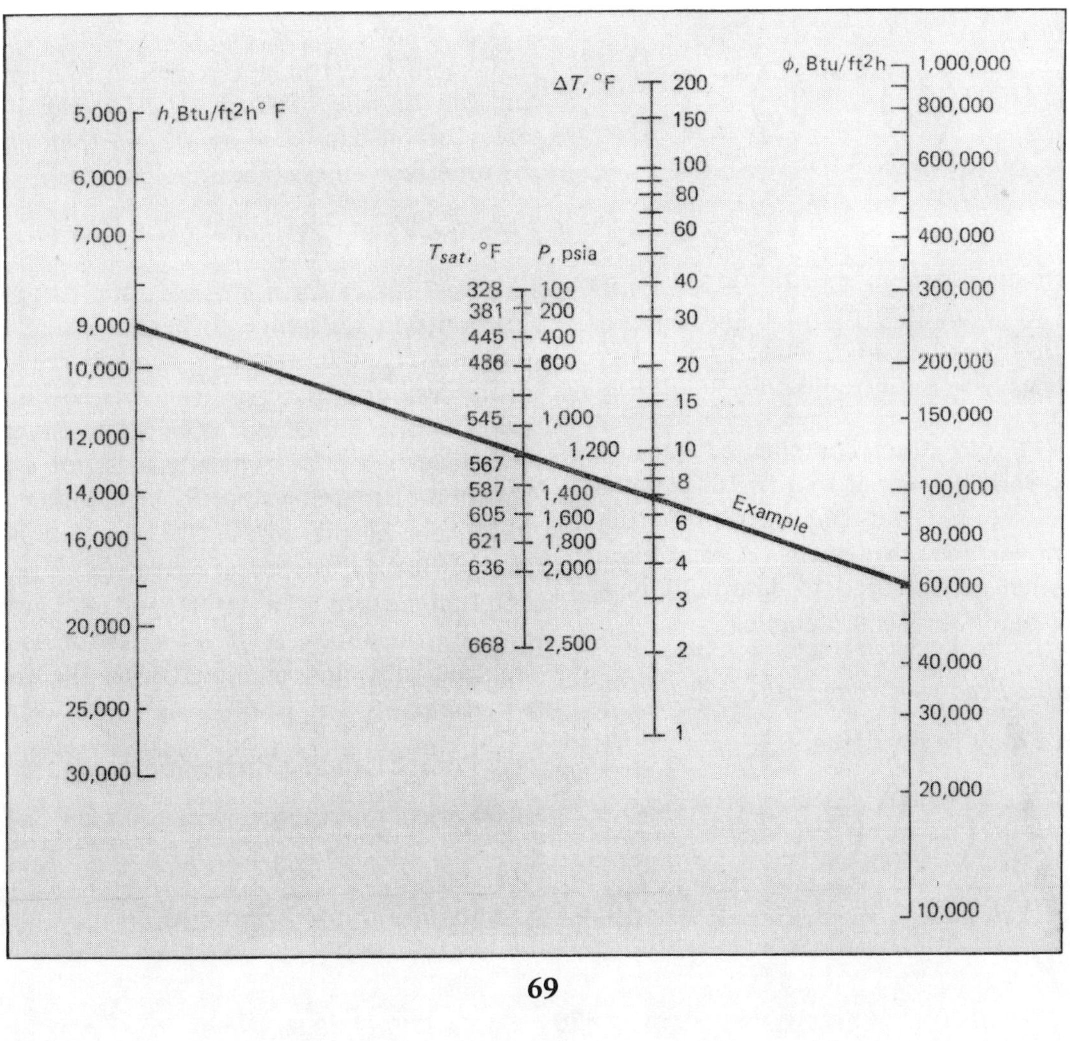

# 11. Charts simplify spiral finned-tube calculations

WHETHER A FINNED TUBE REPRESENTS A SUITABLE choice for an application depends on many factors, such as cost, reliability of the bond between fin and tube, temperature and material limitations, the extent of corrosion and fouling, as well as heat-transfer and pressure-drop requirements.

For calculating heat transfer, the correlation of Robinson and Briggs, and for calculating pressure drop, the correlation of Briggs and Young are often recommended[8] (these equations are for staggered tubes having radial or spiral fins):

$$\frac{H_g d}{12k} = 0.134\left(\frac{Gd}{12\mu}\right)^{0.681}\left(\frac{\mu C_p}{k}\right)^{0.33}\left(\frac{S}{h}\right)^{0.2}\left(\frac{S}{b}\right)^{0.113} \tag{1}$$

$$\Delta p = 18.93\left(\frac{Gd}{12\mu}\right)^{-0.316}\left(\frac{P_t}{d}\right)^{-0.927}\left(\frac{P_t}{P_l}\right)^{0.515}\left(\frac{G^2 N}{g_c \rho}\right) \tag{2}$$

(If performance data on specific finned tubes or bundles are available from a manufacturer, these should be used in preference to the two correlations, which are only suitable for approximate design.)

Equations 1 and 2 are complicated because of the many variables raised to difficult powers, and because thermal and transport properties must be known to solve them. A simplified procedure, such as was used to determine heat transfer over bare tubes, is preferable.[3]

### *Simplifications Lead to Nomographs*

Equation 1 may be rewritten:

$$H_g = 0.295\left(\frac{G^{0.681}}{d^{0.319}}\right)\left(\frac{k^{0.67} C_p{}^{0.33}}{\mu^{0.351}}\right)\left(\frac{S^{0.313}}{h^{0.2}\, b^{0.113}}\right) \tag{3}$$

$$H_g = H_g' F_g \tag{4}$$

$$H_g' = 0.295\,(G^{0.681}/d^{0.319})(k^{0.67} C_p{}^{0.33}/\mu^{0.351}) \tag{5}$$

$$F_g = S^{0.313}/h^{0.2}\, b^{0.113} \tag{6}$$

Fig. 1 solves Equation 5, and Fig. 2, Equation 6.

With finned tubes, the calculation of gas heat-transfer coefficients must take into account the temperature drop along the length of the tube, i.e., fin efficiency. Curves for fin efficiency as a function of fin geometry and outside heat-transfer coefficient have been published.[6,9]. The equation is:

$$\phi = f[\,h(H_g/b\lambda)^{1/2}, Z\,] \tag{7}$$

Fig. 3 solves Equation 7.

Overall heat-transfer coefficients may be expressed on the basis of total extended surface area (or on bare tube area), and then corrected by an effectiveness factor, :

$$H_{go} = [1-(1-\phi)A_f/A_t]\,H_g = \eta H_g \tag{8}$$

Fig. 4 solves for $\eta$, if $A_f$, $A_t$ and $\phi$ are known. The outside resistance to heat transfer, $R_o$, becomes $1/H_{go}$. Values for $A_f$ and $A_t$ are available in tables, and for $A_o$, the area occupied by tubes, which is needed to calculate the gas flow area. (Such a table typically lists, for a particular tube diameter, values for $n$, number of fins/in.; $h$, fin height, in.; and $b$, fin thickness, in.; along with $A_f$, fin surface area, ft²/ft; $A_t$, total external surface area, ft²/ft; and $A_o$, area occupied by the tubes, ft²/ft.) Values for $A_f$, $A_t$ and $A_o$ may also be calculated with the following equations:

$$A_f = (\pi n/24)(4dh + 4h^2 + 2bd + 4bh) \tag{9}$$

$$A_t = (\pi h/24)[n(4dh + 4h^2) + 4nbh + 2d] \tag{10}$$

$$A_o = d/12 + nbh/6 \tag{11}$$

Equation 2 for pressure drop may be simplified as follows:

## Nomenclature

| | |
|---|---|
| $A_f$ | Surface area of fins, $ft^2/ft$ |
| $A_i$ | Inside surface area, $ft^2/ft$ |
| $A_o$ | Area occupied by tubes, $ft^2/ft$ |
| $A_t$ | Total external surface area, $ft^2/ft$ |
| $A_W$ | Wall area, $ft^2/ft$ |
| $b$ | Fin thickness, in. |
| $C_p$ | Specific heat of gas, Btu/(lb)(°F) |
| $d$ | Tube outside dia., in. |
| $d_i$ | Tube inside dia., in. |
| $F_g$ | Geometric factor |
| $F_T$ | Temperature factor for gases, $\mu^{0.316}(460 + T)/M$ |
| $G$ | Gas mass velocity, $lb/(ft^2)(h)$ |
| $h$ | Fin height, in. |
| $H_g$ | Outside gas heat-transfer coefficient, $Btu/(ft^2)(h)(°F)$ |
| $H_{go}$ | Outside gas heat-transfer coefficient corrected for total outside area, $Btu/(ft^2)(h)(°F)$ |
| $h_i$ | Inside heat-transfer coefficient, $Btu/(ft^2)(h)(°F)$ |
| $k$ | Thermal conductivity of gas, $Btu/(ft^2)(h)(°F/ft)$ |
| $M$ | Molecular weight |
| $N$ | Number of tube rows |
| $n$ | Number of fins/in. |
| $\Delta p$ | Gas pressure drop, in. w.c. |
| $P_l$ | Longitudinal pitch, in. |
| $P_t$ | Transverse pitch, in. |
| $R_i$ | Inside resistance to heat transfer, $(ft^2)(h)(°F)/Btu$ |
| $R_o$ | Outside resistance to heat transfer, $(ft^2)(h)(°F)/Btu$ |
| $R_m$ | Metal resistance to heat transfer, $(ft^2)(h)(°F)/Btu$ |
| $S$ | Fin clearance, in. |
| $T$ | Gas temperature, °F |
| $U$ | Overall heat-transfer coefficient, $Btu/(ft^2)(h)(°F)$ |
| $Z$ | $(2h + d)/d$ |
| $\eta$ | Fin effectiveness |
| $\lambda$ | Thermal conductivity of fins, Btu/(ft)(h)(°F) |
| $\mu$ | Viscosity of gas, lb/(ft)(h) |
| $\rho$ | Gas density, $lb/ft^3$ |
| $\phi$ | Fin efficiency |

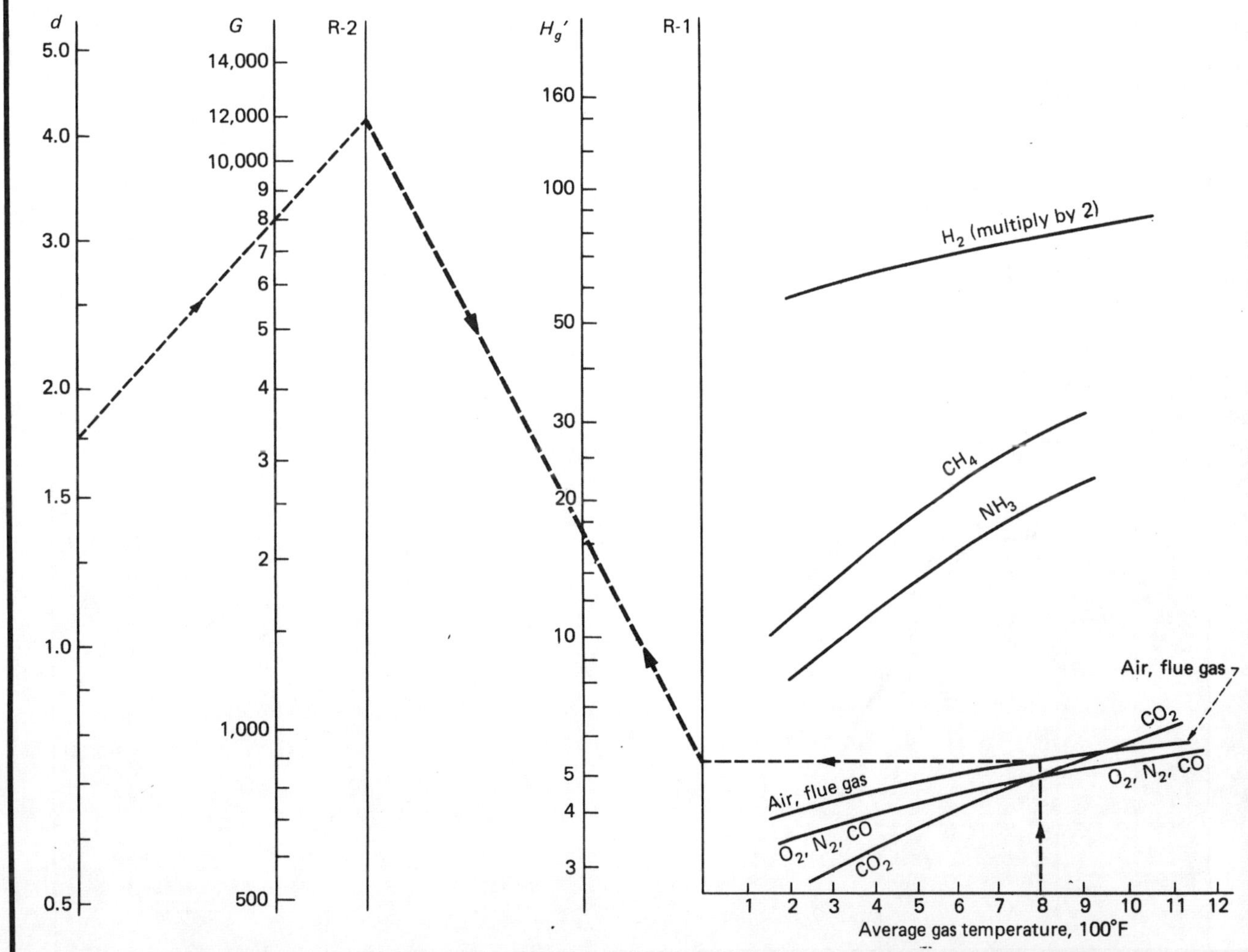


**Fig. 1. Outside heat-transfer coefficient, $H_g$, *is calculated after* $H_g'$, defined by Eq. 5 is determined**

$$\Delta p = 1.58 \times 10^{-8} \left(\frac{G^{1.684} d^{0.611} \mu^{0.316}}{P_t^{0.412} P_l^{0.515}}\right) \times \frac{(T + 460)N}{M} \qquad (12)$$

Fig. 5 solves Equation 12 when $d$, $G$, $T$, $P_t$ and $P_l$ are known. The $\Delta p$ obtained is per row of tubes. Also considered in Fig. 5 are viscosity and molecular weight as functions of temperature.

### *Problem Illustrates Procedure*

Hot air flows at 2.4 million lb/h over a bank of spiral-finned tubes inside a 20-ft × 25-ft duct. The air heats up water flowing inside the tubes.

*Finned-tube details:*

Tube and fin material . . . . . . . . . carbon steel
Tube outside dia, $d$, in. . . . . . . . . . . . . . . . 1.75
Tube inside dia, $d_i$, in. . . . . . . . . . . . . . . . 1.35

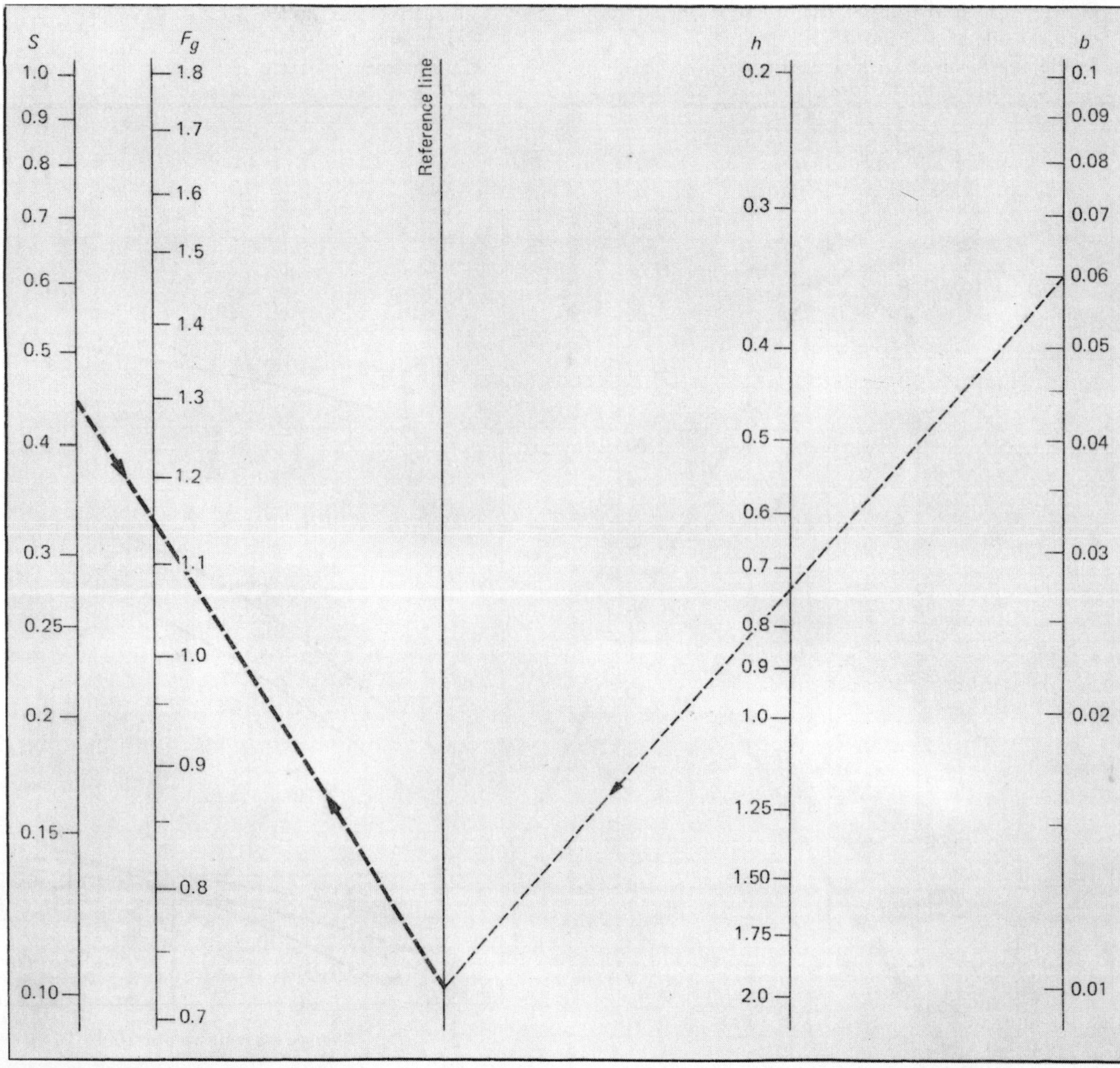


**Fig. 2. Geometric factor, $F_g$, is found, then multiplied by $H_g'$ in Eq. 4 to calculate gas heat-transfer coefficient.**

Fin height, $h$, in. . . . . . . . 0.75
Fin thickness, $b$, in. . . . . 0.06
No. of fins/in. . . . . . . . . . . 2
Fin tube length, ft . . . . . . 24
Transverse pitch, $P_t$, in. . . 4.5
Longitudinal pitch, $P_l$, in. . 4.5
Clearance, $S$, in. . . . . . . . 0.44
No. of tubes across . . . . . . 52

*Thermal data:*

Conductivity of tube and fin, Btu/(ft)(h)(°F) . . . . . . 30
Inlet gas temperature, °F . 900
Outlet gas temperature, °F . . . . . . . . . . . . . . . . 700
Heat transferred, Btu/h . . . . . . . . 120 × $10^6$
Water flow, lb/h . . . . 500,000
Inlet water temperature, °F . . . . . . . . . . . . . . . . 100
Outlet water temperature, °F . . . . . . . . . . . . . . . 340
Log mean temperature difference, °F . . . . . . . 575
Waterside heat-transfer coefficient, Btu/(ft)(h)(°F) . . . . . . . . . . . 1,700

(Because the objective is to present the calculation of the gas heat-transfer coefficient with a finned surface, the calculation of a heat balance and the inside coefficient, $h_i$, are not detailed; the latter is generally high for liquids, so its effect on the overall heat transfer would be less than that of the gas.)

*Calculation procedure:*

1. Area obstructed by tubes, $A_o$ = length of tubes × number of tubes × area obstructed by 1 ft of tube and fins = 24 × 52 × 0.161 = 200 ft$^2$.

Gas free-area = 20 × 25 − 200 = 300 ft$^2$. Gas mass-velocity = 2.4 × $10^6$/300 = 8,000 lb/(ft$^2$)(h).

2. Find $H_{go}$, the outside (gas) heat-transfer coefficient corrected for the total outside area (tube plus fin surface), using Equations 4 and 8, and Fig. 1, 2, 3 and 4, successively. Resistances are subsequently calculated to finally determine the overall heat-transfer coefficient, $U$.

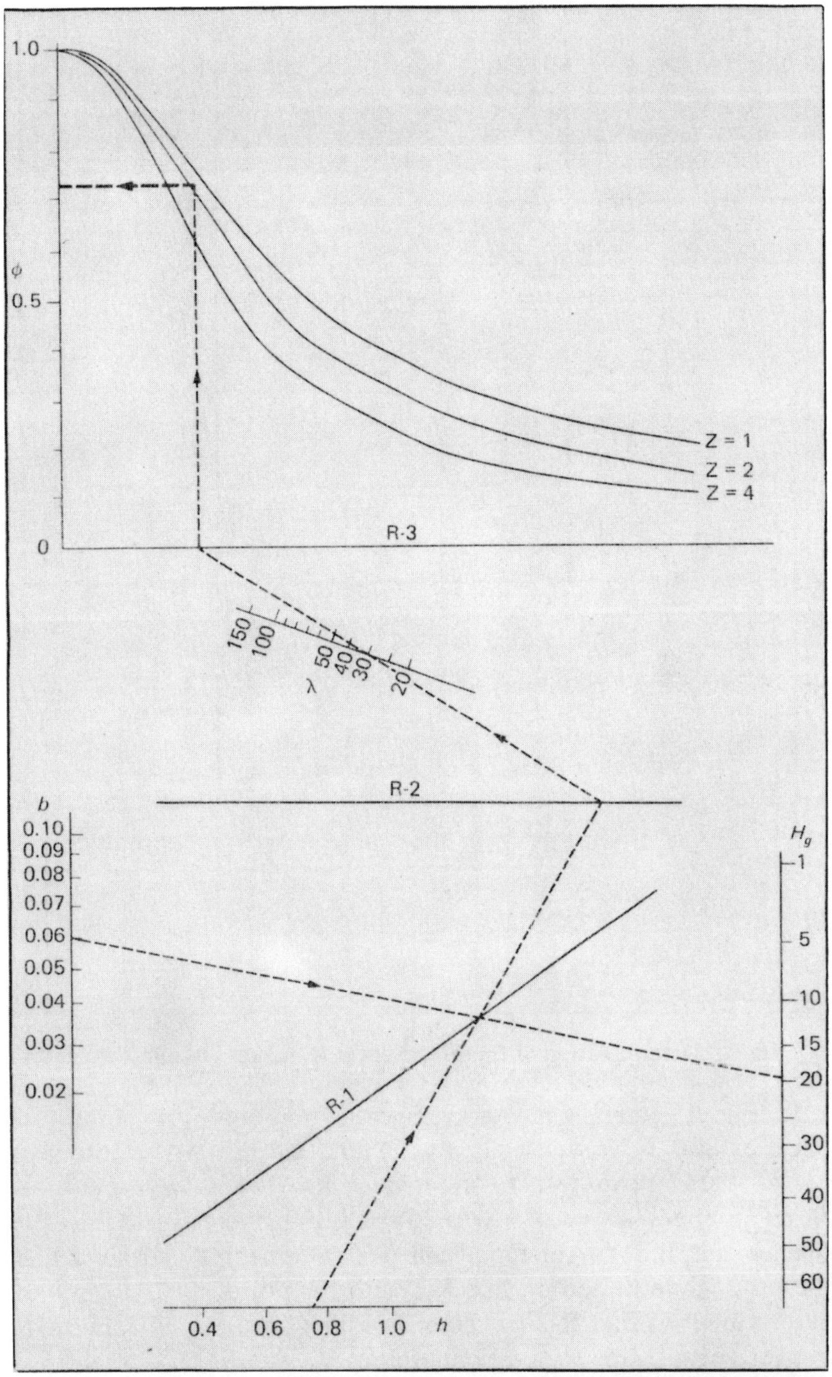


**Fig. 3. Finding fin efficiency, ϕ, leads to establishing fin effectiveness.**

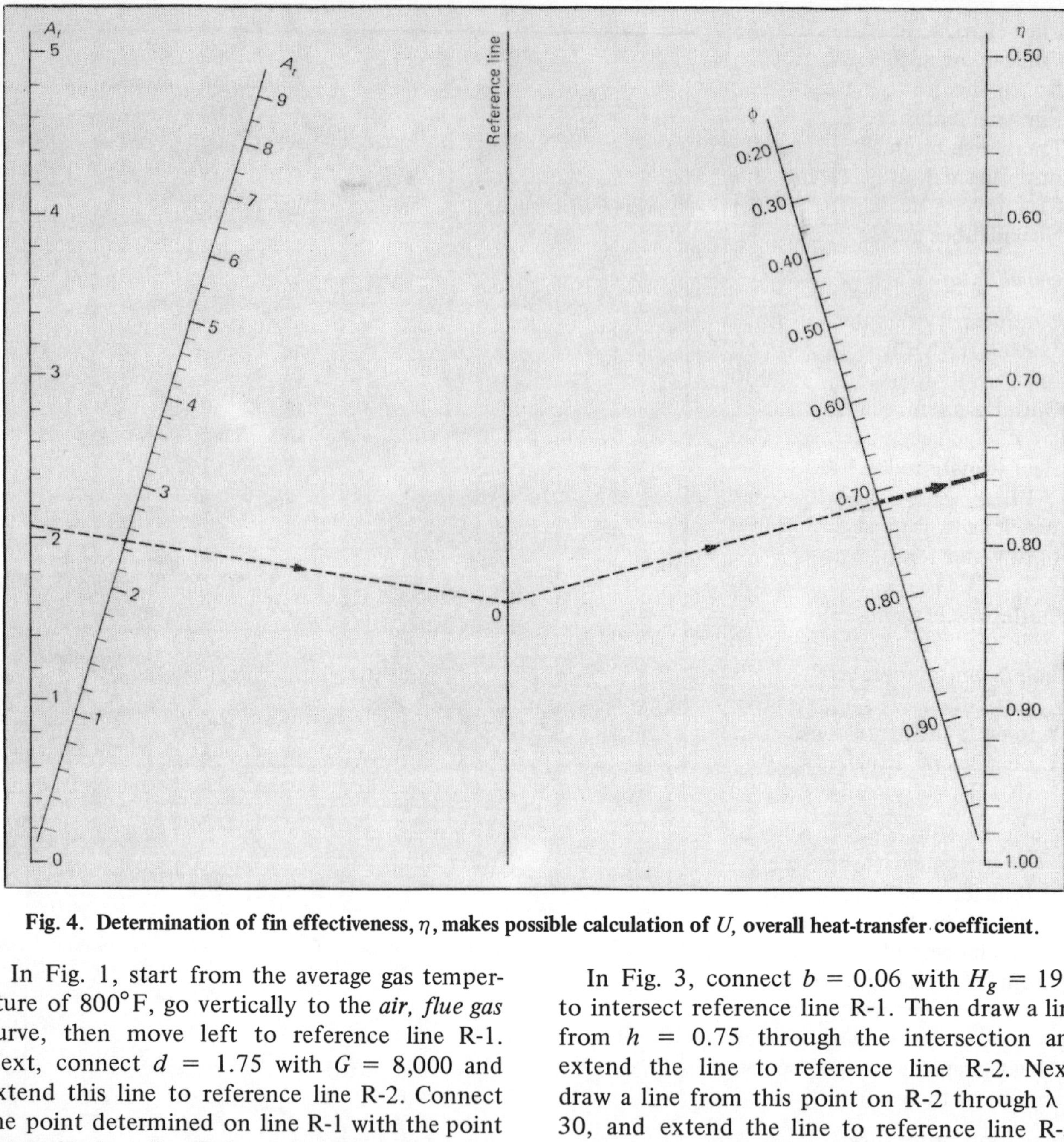


**Fig. 4. Determination of fin effectiveness, $\eta$, makes possible calculation of *U*, overall heat-transfer coefficient.**

In Fig. 1, start from the average gas temperature of 800°F, go vertically to the *air, flue gas* curve, then move left to reference line R-1. Next, connect $d$ = 1.75 with $G$ = 8,000 and extend this line to reference line R-2. Connect the point determined on line R-1 with the point determined on line R-2 to establish a line crossing the vertical line $H_g'$, which gives an $H_g'$ value of 17 Btu/(ft$^2$)(h)(°F).

In Fig. 2, connect $b$ = 0.06 with $h$ = 0.75 and extend the line to the vertical reference line. Connecting the established point on the reference line with $S$ = 0.44 gives $F_g$ = 1.15. Using Equation 4, $H_g$ is then calculated as 19.5 Btu/(ft$^2$)(h)(°F).

In Fig. 3, connect $b$ = 0.06 with $H_g$ = 19.5 to intersect reference line R-1. Then draw a line from $h$ = 0.75 through the intersection and extend the line to reference line R-2. Next, draw a line from this point on R-2 through $\lambda$ = 30, and extend the line to reference line R-3. From this point on R-3, draw a vertical line to $Z$ = 1.85–interpolating between $Z$ = 1 and $Z$ = 2; $Z = (2h + d)/d$. From $Z$ = 1.85, draw a horizontal line to $\phi$ = 0.71.

In Fig. 4, connect $A_f$ = 2.07 with $A_t$ = 2.47, and extend the line to the reference line. From this point on the reference line, draw a line through $\phi$ = 0.71 and extend it to get $\eta$ = 0.76.

Using Equation 8, $H_{go}$ = 0.76 × 19.5 =

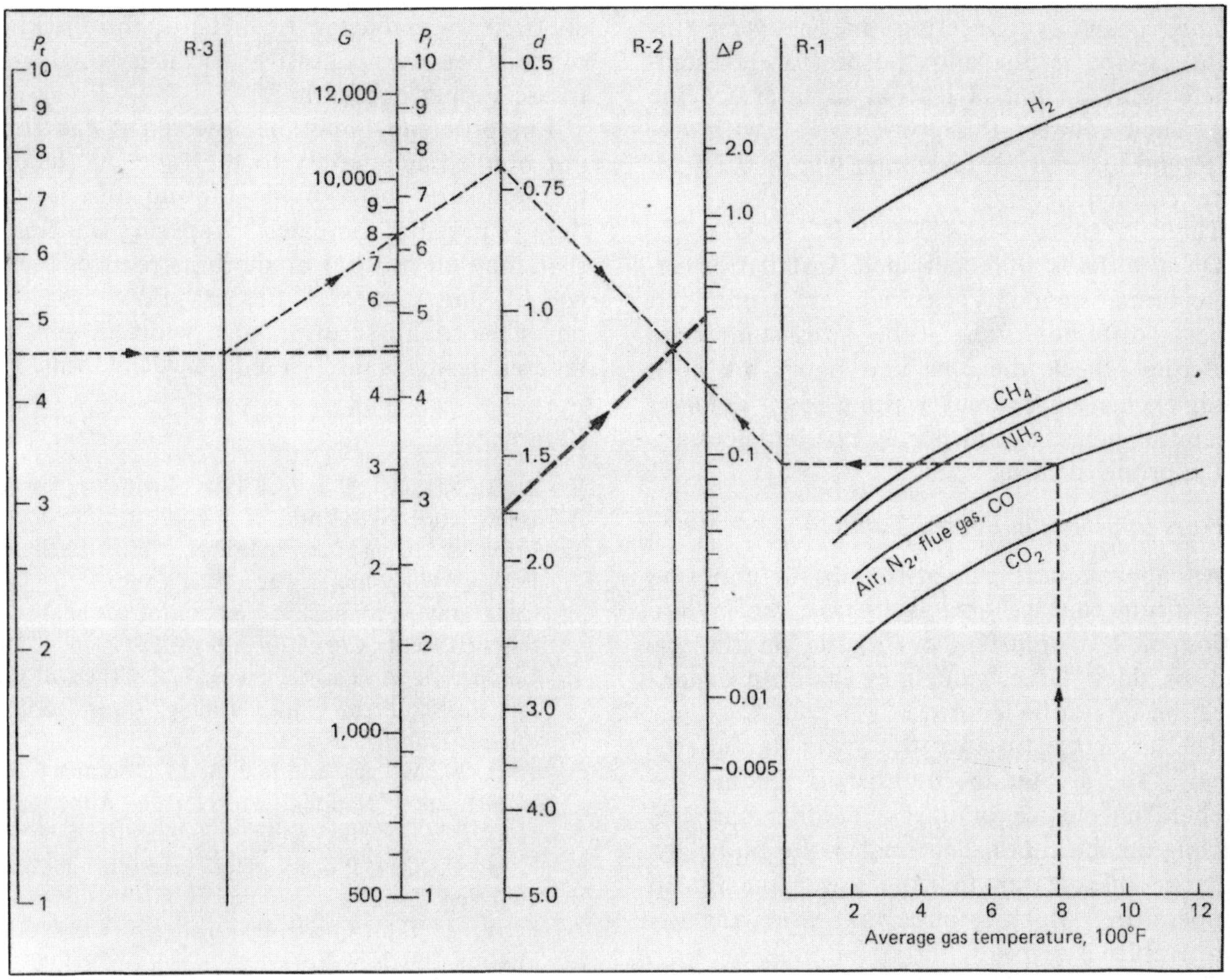


**Fig. 5. Pressure drop found is per row of tubes, so multiply answer by number of rows to obtain total pressure drop.**

14.8 Btu/(ft²)(h)(°F). Calculate the outside resistance: $R_o = 1/H_{go} = 0.0676$ (ft²)(h)(°F)/Btu.

3. The inside coefficient, $h_i$, = 1,700; it may be used to calculate the total outside-area resistance:

$$R_o = \frac{A_t}{h_i A_i} = \frac{12 \times 2.469}{1{,}700 \times \pi \times 1.35}$$

$$= 0.0041 \text{ (ft}^2\text{)(h)(°F)/Btu.}$$

4. Metal resistance is calculated:

$$R_m = \left(\frac{A_t}{A_w}\right)\left(\frac{d}{24\lambda}\right) \ln (d/d_i)$$

$$= 0.00384 \text{ (ft}^2\text{)(h)(°F)/Btu.}$$

5. The overall heat-transfer coefficient, $U$, is:

$$U = 1/(R_o + R_i + R_m) = 13.2 \text{ Btu/(ft}^2\text{)(h)(°F).}$$

(Fouling may be taken into account in the foregoing calculation.)

6. The surface area required = (120 × 10⁶)/(575) (13.2) = 15,810 ft². The number of tubes deep in the bundle = 15,810/(2.469)(24)(52) = 5.1 (use 6).

7. To find the gas pressure drop, use Fig. 5. Starting from $T$ = 800°F, draw a vertical line up to the *air, flue gas* curve, then move horizontally to reference line R-1. Next, draw a line from $P_t$ = 4.5 to $P_l$ = 4.5 through R-3, then connect this point with a line through $G$ = 8,000 to the

*d* line. Draw a connecting line between this point on the *d* line and the previously established point on line R-1 so as to intersect line R-2. Then connect this point on R-2 with $d = 1.75$ and extend it to obtain $\Delta p = 0.38$ in. water column.

The total gas $\Delta p = 6 \times 0.38 = 2.2$ in. w.c.

On the basis of experience, installation requirements, economy desired, etc., one can choose different finned-tube sizes and mass velocities, check the $\Delta p$s, and figure the optimum exchanger without waiting for a detailed computer analysis, which may not be necessary in the proposal stage.

### Factors in Selecting Finned Tubes

An approximate rule-of-thumb for choosing finned-tube surface area is to take the inverse ratios of heat-transfer coefficients on the gas and air sides. Other guidelines based on experience should not be ignored.[1,6]

The larger the fin area, the lower the fin efficiency. The greater the number of fins/in., the likelier the chance of dust and other particles settling on the tubes. The smaller the fin height and the thicker the fin, the lower the fin-tip temperature, which is important when the gas temperature is high.

For applications in which the gas heat-transfer coefficient is low, the thermal conductivity of steel is good enough to allow adequate fin efficiency to be obtained with a reasonable fin thickness.[2] With higher fin-side heat-transfer coefficients and large fin heights, the thickness of steel becomes excessive, and it is worthwhile to use copper or aluminum.

Fin efficiency may be lowered if the fin is not bonded adequately to the tube. As the contact pressure between the fin and tube lessens, the gas pressure increases.[4] Tapering the fins so that they are thicker at the base reduces the fin weight and increases the gas flow area, but raises the manufacturing cost, which is why the tapered design is rarely used for welded fins.

### References


1. Boyen, J. L., "Practical Heat Recovery," Wiley-Interscience, New York.
2. Fraas, A. P. and Ozisik, M. N., "Heat Exchanger Design," Wiley-Interscience, New York.
3. Ganapathy, V., Quick estimation of gas heat-transfer coefficients, *Chem. Eng.,* Sept. 13, 1976, p. 199.
4. Gardner, K. A., and Carnavos, T. C., Thermal Contact Resistance in Finned Tubing, *Trans. ASME J. of Heat Transfer,* Nov. 1960.
5. Kays, W. M. and London, A. L., "Compact Heat Exchangers," National Press, Palo Alto, Calif., 1955.
6. Kern, D. Q., "Process Heat Transfer," McGraw-Hill, New York.
7. Perry, J. H., "Chemical Engineers' Handbook," McGraw-Hill, New York.
8. Rohsenow, W. M. and Hartnett, J. P., "Handbook of Heat Transfer," McGraw-Hill, New York.
9. Kern, D. Q. and Kraus, A. D., "Extended Surface Heat Transfer," McGraw-Hill, New York, 1972, pp. 101-104.

# 12. Charts aid quick estimate of gas heat-transfer coefficient for finned tubes

FINNED TUBES ARE USED EXTENSIVELY IN conventional and waste-heat boilers, fired heaters, and exchangers. The choice of finned-tube size and arrangement is governed by several factors. To consider the various alternatives, it is first necessary to evaluate the gas-side film coefficient for the finned tubes.

The most widely used equation is that of Robinson and Briggs for heat transfer. This equation is quite involved, however, and requires knowledge of gas properties. The nomographs included here give the gas-side heat-transfer coefficient, and also an overall fin-effectiveness factor, which takes into account fin efficiency (a function of the temperature variation along the length of the fin) and the ratio of fin area to total extended outside area. These charts are based on a metal thermal conductivity of 30 Btu/(sq ft)(hr)(deg F/ft), and fin heights of 0.60 and 0.75 in.

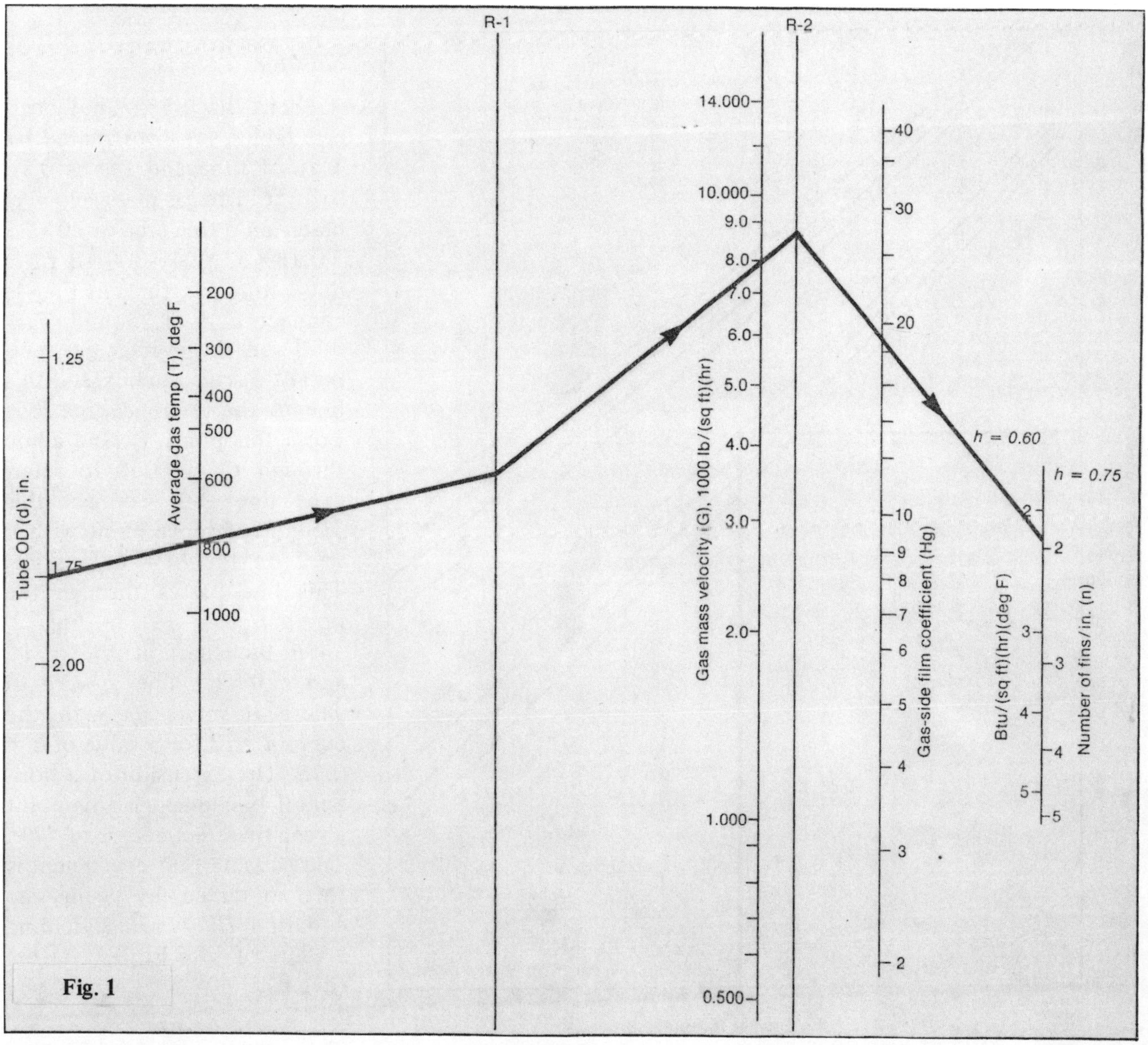


Fig. 1

Once the gas-side corrected film coefficient is known, it can be used to solve the following equation:

$$U = 1/(1/H_g + r_0 + r_i(A_0/A_i) + R_{wall} + A_0/A_iH_t)$$

where $U$ is the overall heat-transfer coefficient based on the total extended outside area, $H_g$ is the corrected gas-side film coefficient, $H_t$ is the tube-side film coefficient, and $r_0$ and $r_i$ fouling resistances for the outside and inside surfaces, all in units of Btu/(sq ft)(hr)(deg F), $A_0$ and $A_i$ are the outside and inside surface areas in sq ft and $R_{wall}$ is the resistance of the tube wall referred to the total extended surface in Btu/(sq ft)(hr)(deg F).

### *Example*

Calculate the gas-side film coefficient for 2.4 million lb/hr of hot gases from a turbine exhaust flowing over a finned-tube bank in a 20X25-ft duct. The gases are cooled from 900F to 700F. The tubes are 1.75 in. O.D., fin thickness is 0.06 in., fin height is 0.75 in., and there are 2 fins per in. Tube length is 24 ft, and horizontal tube width is 52; a transverse and longitudinal pitch of 4.5 in. is assumed.

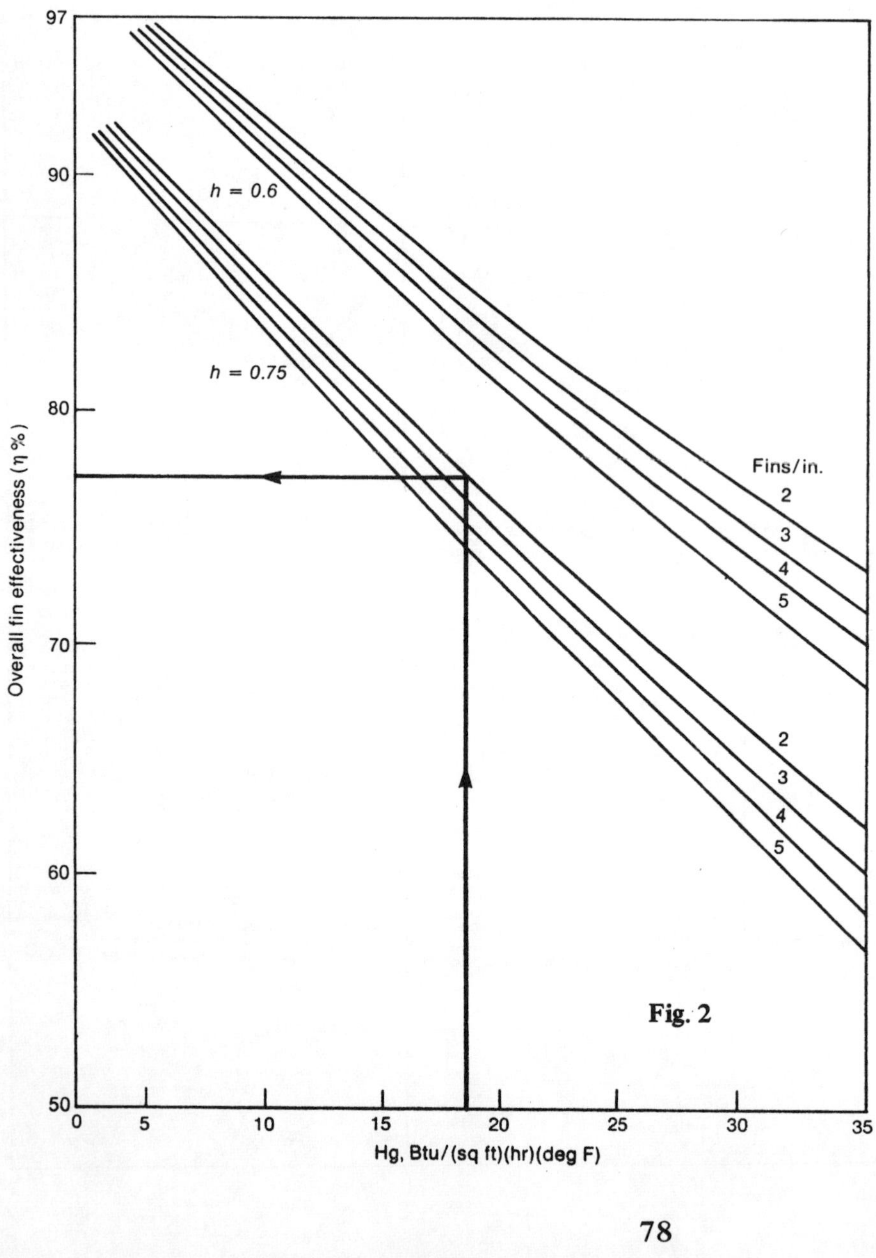


**Fig. 2**

### *Solution*

From the tube manufacturers' tables, area obstructed by 1 ft of tube and fins is 0.16 sq ft/ft. The gas mass velocity, based on a free area of 20X25 −0.16X52X24=300 sq ft, is 8000 lb/(sq ft)(hr). In Fig. 1, connect $d$ = 1.75 and $T$ = 800F on the average gas temperature scale, and extend this line to the reference line R-1. From this point, extend a line through $G$ = 8000 to reference line R-2. Connect this second reference point with $n$ = 2 on the $h$ = 0.75 scale. This line passes through the point $H_g'$ = 19. In Fig. 2, enter the chart at $H_g'$ = 19 and extend a line upward to where it intersects with the curve $n$ = 2 for a value of $h$ = 0.75. The extension of a horizontal line through this point gives a fin effectiveness of 77%. The gas-side film coefficient is then obtained by using the equation $H_g = \eta H_g'$, yielding a value of 14.6 Btu/(sq ft)(hr)(deg F).

# 13. Charts help estimate nonluminous heat-transfer coefficient in tube bundles

***Detailed combustion calculations for fuels and Hottel's charts are not needed when these nomographs are used to estimate nonluminous radiation heat-transfer coefficients between hot gases and tube bundles.***

A NUMBER OF FACTORS ENTER INTO THE CALculation of nonluminous radiation heat-transfer coefficients between hot gases and tube bundles.

Net interchange of radiation between gases and tube bundles is given by the basic equation.

$$Q/A = \sigma[\epsilon_g T_g{}^4 - \alpha_g T_o{}^4] \qquad (1)$$

For quick practical estimates, a good assumption that simplifies the above equation is $\epsilon_g = \alpha_g$. Then

$$Q/A = \epsilon_g\, \sigma[T_g{}^4 - T_o{}^4] \qquad (2)$$

From this, the nonluminous heat-transfer coefficient $h_N$ may be obtained as follows.

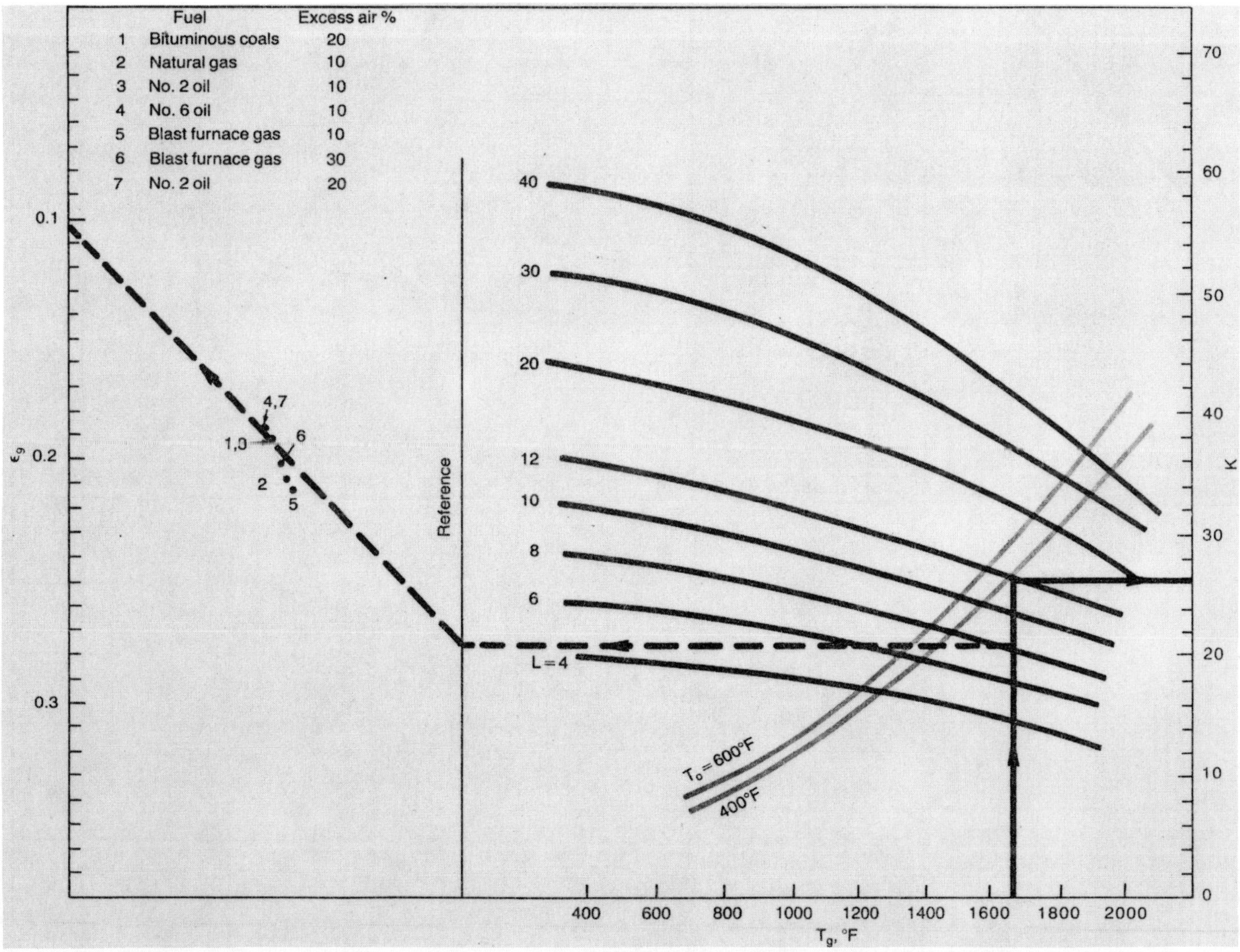


**Fig. 1. Nomograph for determining $h_N$**

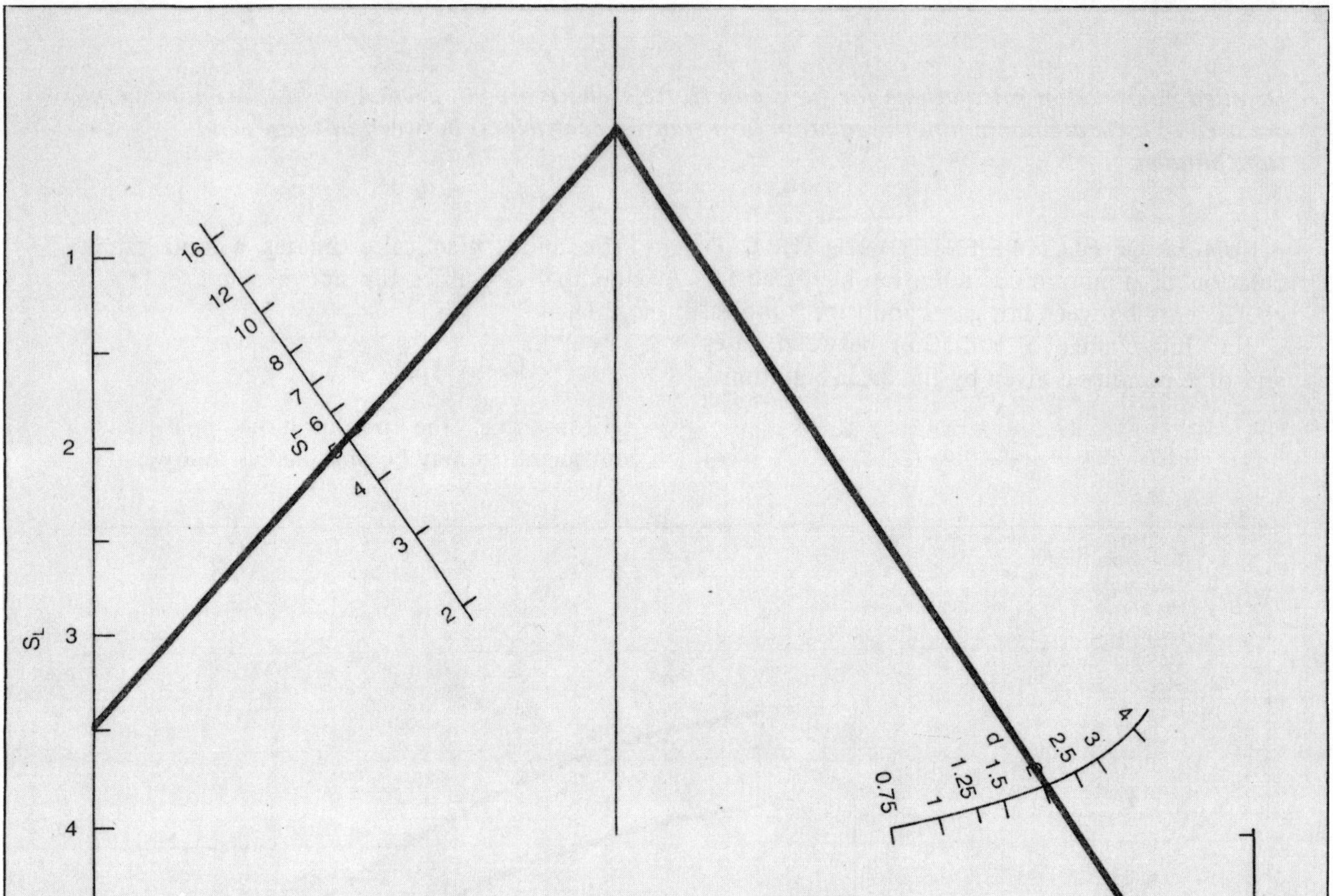


**Fig. 2. Bundle beam length determination**

$$h_N = \frac{Q}{A(T_g - T_o)} = \epsilon_g \sigma \frac{[T_g^4 - T_o^4]}{(T_g - T_o)} \quad (3)$$

Where

$$\epsilon_g = \epsilon_c + \eta\epsilon_w - \Delta\epsilon \quad (4)$$

The emissivity of hot flue gases is dependent upon the emissivity of carbon dioxide, $\epsilon_c$, emissivity due to water vapor, $\epsilon_w$ and correction terms $\eta$ and $\Delta\epsilon$[1].

These values may be obtained from the well known Hottels' charts, once the partial pressures of carbon dioxide and water vapor, the beam length of the heat receiving surface and the gas temperature are known.

The beam length, $L$, of a tube bundle is given by:

$$L = 1.08(S_T S_L - 0.785 d^2)/d \quad (5)$$

Fig. 1 gives $L$ if $S_T$, $S_L$ and $d$ are known.

Hence it may be noted that an estimate of $h_N$ requires information on partial pressures of $CO_2$ and $H_2O$ in flue gas, which in turn implies availability of results of combustion calculations of fuels. This is a lengthy procedure.

However, a simplified procedure has been developed based on detailed calculations with various fuels. Equation 3 is rewritten as:

$$h_N = K\epsilon_g \quad (6)$$

$\epsilon_g$ is calculated for various fuels at normally used excess air factors and presented in Fig. 2 along with $K$. The accuracy of this figure has been checked through detailed calculations and a maximum error of 5 percent was observed. Hence, for quick engineering estimates of $h_N$, Figs. 1 and 2 may be used.

The advantages are:

- Detailed combustion calculations for fuels need not be carried out and thus a lot of time is saved (sometimes even the fuel analysis may not be readily available).
- Reference to Hottel's charts is not necessary, again saving time.
- Quick comparisons of $\epsilon_g$ as a function of fuel used may be made. This is an important factor as furnaces of boilers, fired heaters etc. have to fire different fuels during their operation, and performance with alternative fuels must be known.

An example is worked out to illustrate the versatility of the charts.

### *Example*

A boiler is fired with No. 2 oil at 10 percent excess air. Determine the nonluminous radiation heat-transfer coefficient over a tube bank when the average flue gas temperature is 1650°F and tube wall temperature is 400°F.

$S_T = 5.0$ in., $S_L = 3.5$ in. and $d = 2.0$ in.

### *Solution*

In Fig. 1, connect $S_L = 3.5$ with $S_T = 5$ and extend to cut line 1. Connect with $d = 2$ and extend to cut $L$ scale at 7.8 in.

In Fig. 2 go up from $t_g = 1650$°F to cut $t_o = 400$°F and move right to cut $K$ scale at 26. Extend the vertical from $t_g = 1650$ to cut $L = 7.8$ and move left to cut the reference line. Connect with point 3 (No. 2 oil) and extend to cut $\epsilon_g$ scale at 0.105.

Hence $h_N = 0.105(26) = 2.74$ Btu/ft²h°F.

### *Nomenclature*

| | |
|---|---|
| $A$ | Surface area, ft² |
| $d$ | Tube diameter, in |
| $\alpha_g$, $\epsilon_g$ | Absorptivity and emissivity of gas |
| $h_N$ | Non-luminous radiation heat transfer coefficient, Btu/ft²h°F |
| $\epsilon_c$, $\epsilon_w$ | Emissivity due to carbon dioxide and water vapor |
| $\eta$, $\Delta\epsilon$ | Correction terms used in Eq. 4 |
| $K$ | Term equal to $\sigma\left(\frac{T_g^4 - T_o^4}{T_g - T_o}\right)$ Btu/ft²h°F |
| $L$ | Beam length, in |
| $Q$ | Heat duty, Btu/h |
| $\sigma$ | Steffan Boltzman constant for radiation |
| $S_T$, $S_L$ | Transverse and longitudinal pitch of tubes, in |
| $T_g$, $T_o$ | Absolute temperatures of gas and tube wall, °R |
| $t_g$, $t_o$ | Temperatures of gas and wall, °F |

### *Reference*

1. McAdams, W. H., "Heat Transmission," 3rd Ed. McGraw-Hill.

# 14. Nomogram aids sizing of piping insulation

IN MANY PIPING INSULATION PROBLEMS, calculation of the resistance to heat transfer for the insulation becomes a tedious problem. This resistance is defined as:

$$R = D \ln(D/d)/2K$$

and the heat loss is:

$$Q = \Delta T/\Sigma(d \ln(D/d)2k) = \Delta T/\Sigma R$$

where:

- $R$ = resistance to heat transfer, (h)(ft$^2$)(°F)/Btu
- $D$ = outside diameter of insulation, in.
- $d$ = inside diameter of insulation, or diameter of the pipe, in.
- $k$ = thermal conductivity of the insulation, Btu/(h)(ft$^2$)(°F/in.)
- $\Delta T$ = the temperature difference across the insulation, °F.

This chart relates $R$, $D$, $d$, and $k$, so that any one of the four can be determined in terms of the other three. It thus eliminates a trial-and-error calculation.

### *Example*

If a 24-in. pipe is to be insulated with material having a thermal conductivity of 1.0 Btu/(h)(ft$^2$)(°F/in.) so that the overall resistance of the insulation to heat loss from the hot pipe to the ambient air is 5.0 (h)(ft$^2$)(°F)/Btu, what are the thickness and outside diameter of the insulation? On the chart, connect $R$ = 5.0 with $k$ = 1.0 and extend the straight line to its intersection with the reference line. Connect this intersection with $(d/2)$ = 12, and read off $(D/2)$ = 16.2 in. Thus the thickness is (16.2 − 12) = 4.2 in.

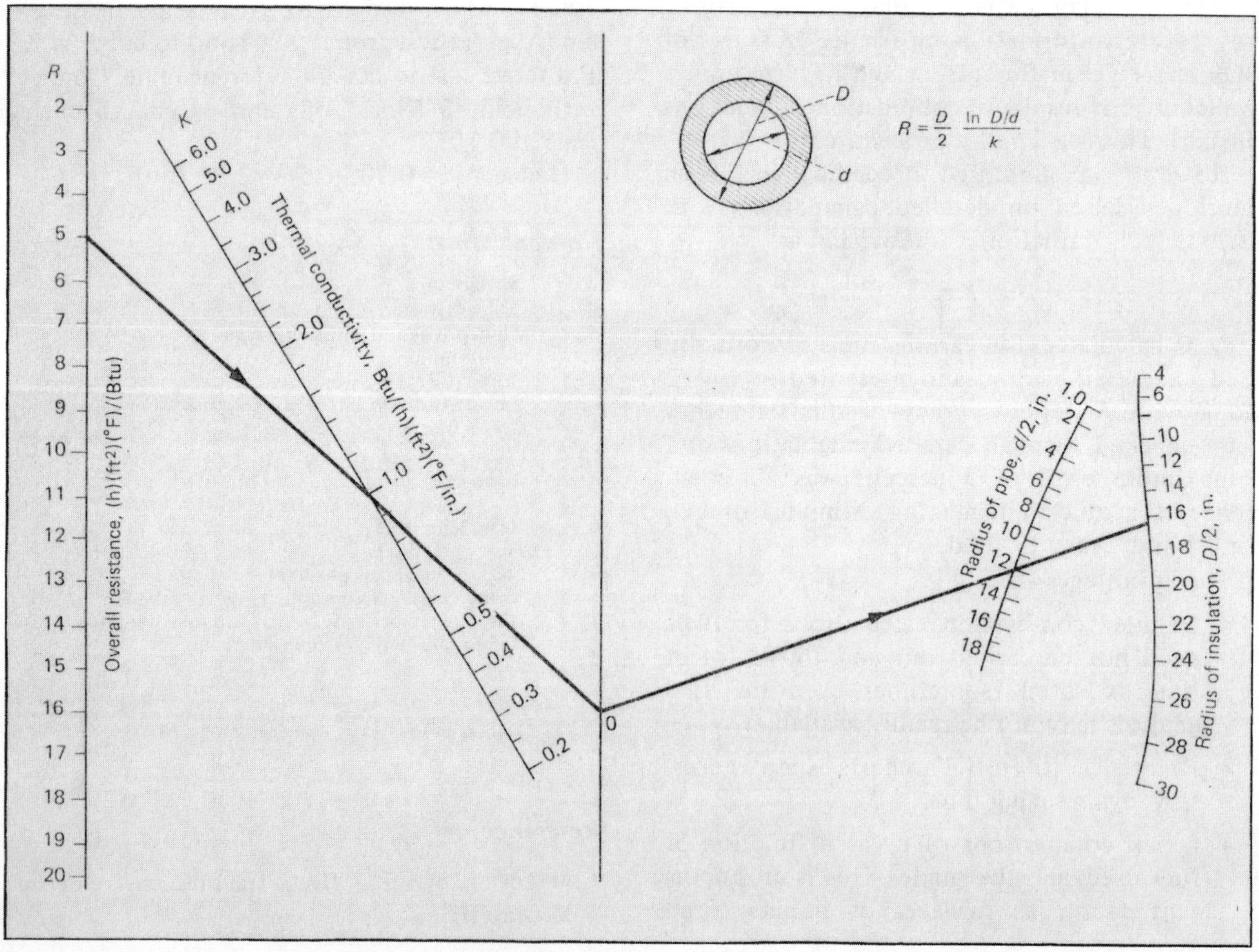

# 15. Chart estimates insulation thickness and heat loss

THE HEAT LOSS FROM INSULATED PIPING IS AN important factor affecting the energy economics of process plants. Plant engineers should have an idea of the heat loss, which is a function of several variables:

$$\alpha = \frac{(t_1 - t_a)}{\left[\left(r^2 \ln \frac{r_2}{r_1}\right)/k + 1/f\right]} \qquad (1)$$

Often the insulation thickness, $t$, is decided by limiting the heat loss to a particular value. Then solving for $t$ $(r_2 - r_1)$ becomes difficult because the expression in the denominator involves $r_1$ and $t$. Estimating $Q$ given $r_1$ and $t$ is less difficult.

However, both types of problems can be solved by the accompanying chart, which takes into account the effect of outside film coefficient $f$. This coefficient generally varies from 1.5 to 2.5 Btu/ft$^2$-hr-°F. Table 1 gives $f$ values for typical situations.

The surface temperature $t_s$ is obtained from:

$$(t_s - t_a) = Q/f \qquad (2)$$

**Table 1. Values of $f$***

| | |
|---|---|
| Still air | 1.2–1.8 |
| 7.5 mph wind | 2.0–4.0 |
| 15 mph wind | 2.5–5.0 |

*Btu/ft$^2$-hr-°F. For preliminary estimates, use 2.0.

### *Example*

Determine the thickness of insulation to limit heat loss to 60 Btu/ft$^2$-hr in a 3-in. NPS pipe. Pipe temperature $t_1$ is 580°F; $t_a$, the ambient temperature, is 80°F. Insulation $K$ value is 0.5 Btu/ft$^2$-hr-°F/in. and outside film coefficient is 2.0 Btu/ft$^2$-hr-°F. What is the surface temperature of the insulation?

### *Solution*

For the solution, connect $(t_1 - t_a) = 500$ with $Q = 60$ and extend to cut line 1 (dashed lines) at A. Connect $f = 2.0$ with point A and extend to cut line 2 at B. Connect B with $K = 0.5$ to cut line 3 at C. The horizontal from C and the vertical from pipe size 3 intersect the curve corresponding to $t = 2.5$ in. Hence 2.5, or the next standard size of insulation.

From Equation 2, $(t_s - t_a) = 60/2 = 30$. Hence $t_s = 30 + 80 = 110$°F.

The effect of using a different thickness of insulation and the corresponding heat loss can easily be calculated. For example, using 2.0-in. thick insulation, we see that (solid lines) $Q = 78$ Btu/ft$^2$-hr and surface temperature increases to $(78/2 + 80) = 119$°F.

### *Life Cycle Costing*

Life cycle costing is a method of evaluating the cost of owning and operating the system over the life of the equipment.

The following expressions aid evaluation of the life cycle cost:

$$TCC = CCO + CC \qquad (3)$$

where:

$$CCO = C_a \left(\frac{1+e}{1+i}\right) \left[\frac{1 - \left(\frac{1+e}{1+i}\right)^N}{1 - \left(\frac{1+e}{1+i}\right)}\right] \qquad (4)$$

The alternative with the lowest $TCC$ is the best, as shown by the following example.

### *Example*

In choosing between the 2.0 and 2.5-in. thickness of insulation, the plant engineer has the following data:

Length of pipe – 1,000 ft
Cost of energy – \$4/MMBtu

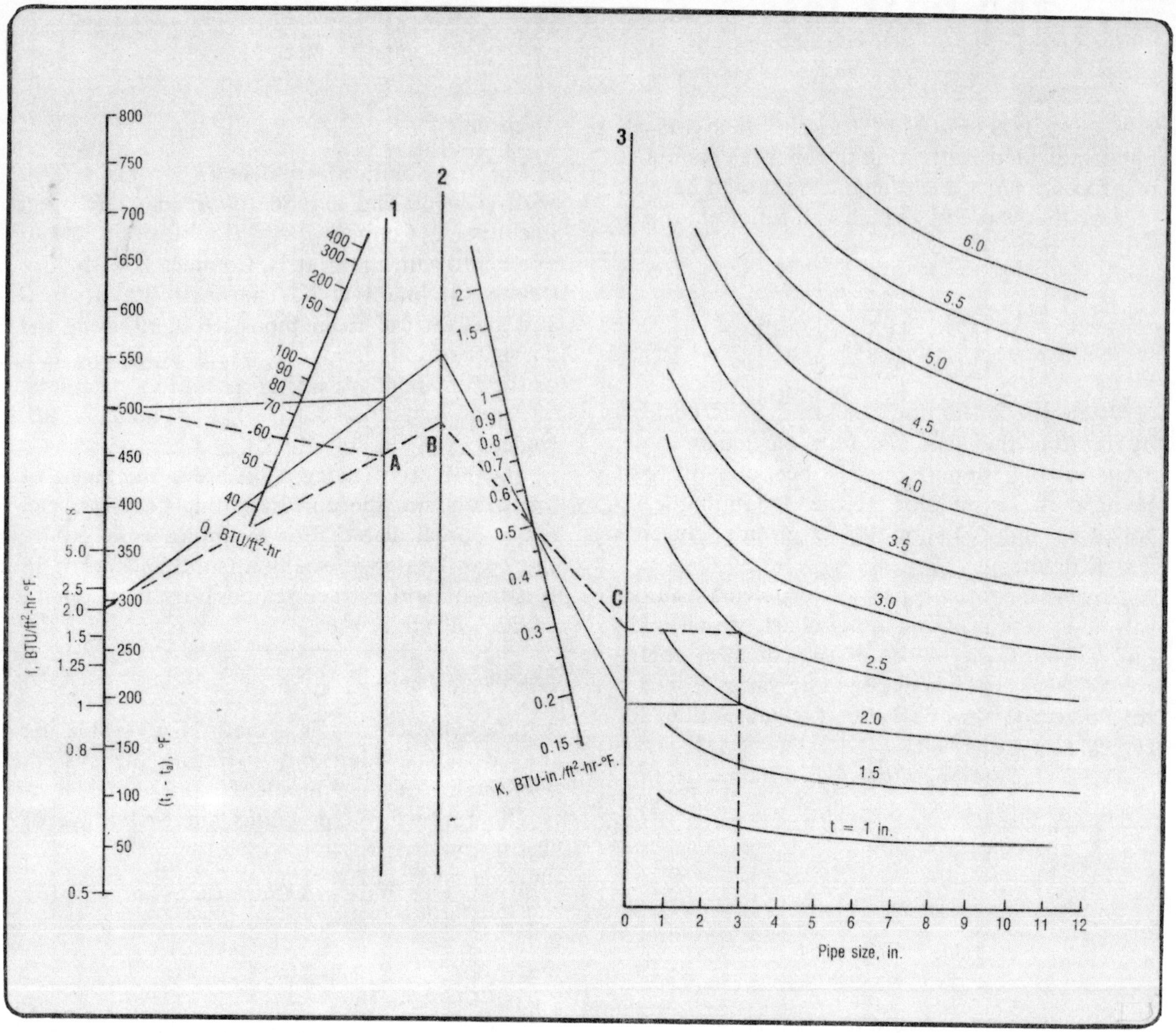


Interest and escalation factors are 0.15 and 0.08 (15% and 8%).

Cost of insulation for 2.5-in. insulation including labor is \$8,000; for 2.0 in. it is \$5,500.

Annual period of operation is 8,000 hours. Life of the equipment is 20 years.

For the solution, calculate the annual dollar value of the heat loss, $C_a$:

$$C_a = N\pi\left(\frac{d+2t}{12}\right)L\alpha C_e/10^6$$

For 2.5-in. thickness:

$$C_a = \pi\left(\frac{3.5+5}{12}\right) \times 1{,}000 \times 60 \times 4 \times 8{,}000/10^6 = \$4{,}256$$

Capitalized cost of operation, $CCO$:

$$CCO = 4{,}256 \times \frac{1.08}{1.15}\left[\frac{1-\left(\frac{1.08}{1.15}\right)^{20}}{1-\left(\frac{1.08}{1.15}\right)}\right] = \$47{,}062$$

Total capitalized cost, $TCC = 47{,}062 + 8{,}000 = \$55{,}062$. Similarly, for the 2.0-in. thickness, $TCC = \$59{,}639$.

The 2.5-in. thickness is better in the long run, though its initial cost is higher.

If the plant runs for only 1,000 hr during the course of a year, the same calculations yield:

$$TCC = \$13{,}883 \text{ for 2.5 in., and } \$12{,}226 \text{ for 2.0 in.}$$

When the period of operation is less, a lower investment is adequate.

In this way, alternatives may be evaluated without elaborate studies.

### Bibliography

Ganapathy, V. Applied Heat Transfer, PennWell Publishing Co., 1982.

### Nomenclature

| | |
|---|---|
| $C_a$ | Annual cost of operation, \$ |
| $CC$ | Fixed or capital cost, \$ |
| $C_e$ | Cost of energy, \$/MMBtu |
| $CCO$ | Capitalized cost of operation, \$ |
| $e$ | Escalation factor |
| $i$ | Interest rate |
| $N$ | Life of the plant in years |
| $f$ | Outside film coefficient, Btu/ft²-hr-°F |
| $Q$ | Heat loss, Btu/ft²-hr |
| $t_a, t_s, t_1$ | Ambient, surface and pipe temperatures, °F |
| $TCC$ | Total capitalized cost, \$ |
| $K$ | Thermal conductivity of insulation, Btu/ft²-hr-°F/in. |
| $L$ | Length of pipe, ft |

# 16. Chart relates heat loss and surface temperature

IN MANY PROBLEMS INVOLVING ROOM heaters and the selection of insulation and refractory, one needs to calculate the surface temperature of a pipe that will give a certain rate of heat emission, or vice versa. The relation, including both convective and radiant heat, is

$$Q = 0.296(T_s - T_a)^{1.25} + 0.174e[(T_s/100)^4 - (T_a/100)^4]$$

Where:

$Q$ = heat loss, Btu/(hr)(ft$^2$)
$T_s$ = surface temperature, °R
$T_a$ = ambient temperature, °R
$e$ = surface emittance, ratio to a black-body

This equation is not too difficult to solve for $Q$, when the surface temperature is known, but it requires trial and error when $Q$ is known and the surface temperature is sought. Trial and error is time-consuming, even with a pocket calculator. The accompany nomograph solves the equation in either direction without the need for trial and error.

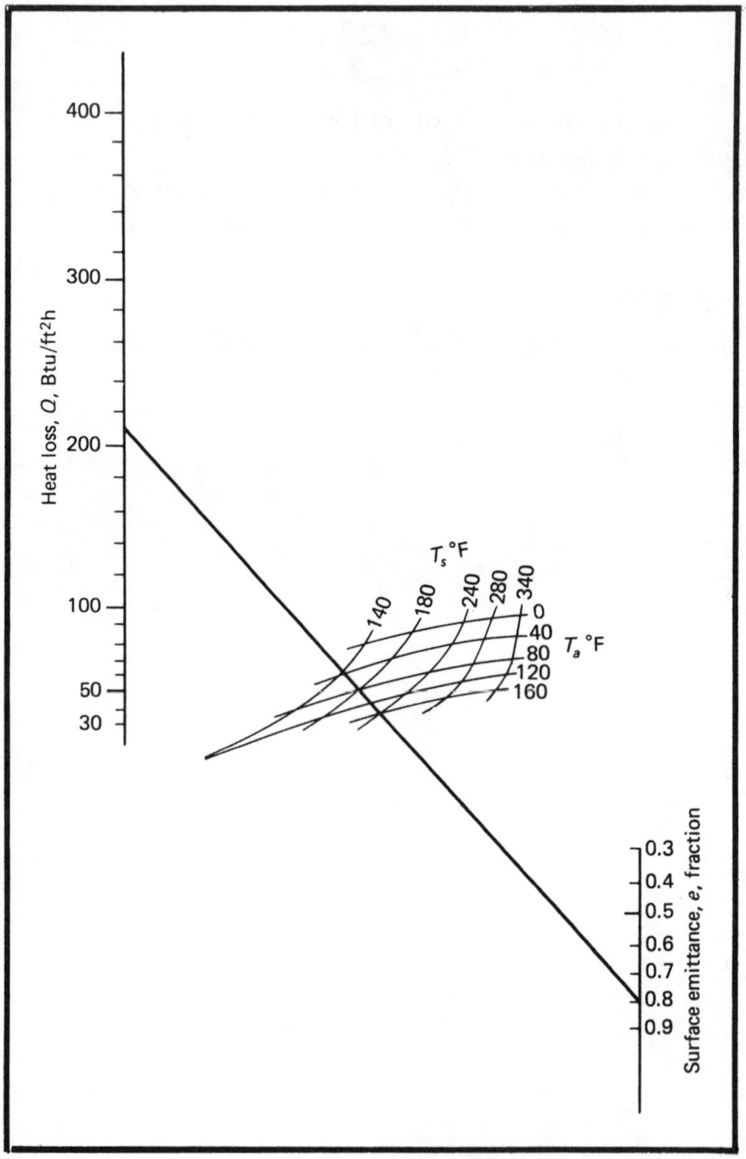


### *Example*

Given a total heat-release rate of 210 Btu/(hr)(ft$^2$), and ambient temperature of 80°F, and an emissivity of 0.8, what is the surface temperature of the pipe?

### *Solution*

On the chart, connect $Q$ = 210 with $e$ = 0.8; note the intersection of this straight line with the curved line for $T_a$ = 80, and read $T_s$ = 180°F.

# 17. Chart eases heat-loss calculations

A SURFACE AT A TEMPERATURE $T_1$, WHEN INsulated, loses heat through the insulation to the atmosphere which is at ambient temperature $T_a$. The outer surface of the insulation is at temperature $T_s$.

Under steady-state conditions, the following equations govern:

$$Q = (T_1 - T_s)/(L/K) = f(T_s, T_a, e) \quad (1)$$

where:

$Q$ = heat loss, Btu/ft²-hr

$T_1$, $T_a$, and $T_s$ are as previously defined, °F

$L$ = insulation thickness, in.

$K$ = thermal conductivity of insulation, Btu-in./ft²-hr-°F

$e$ = emissivity of the surface

$f$ = a complicated function involving radiation between the outer surface and the atmosphere and the emissivity of the surface:

$$= 0.296(T_s - T_a)^{1.25} + 0.174e\,[(T_s/100)^4 - (T_a/100)^4] \quad (2)$$

Usually heat loss $Q$ is fixed by experience and the insulation outer surface temperature $T_s$ is solved for, which is tedious. The ratio $L/K$ is then solved for.

With the chart, one can arrive at the values of $L/K$ and $T_s$. Even if $L/K$ is given and $Q$ is desired, one or two trials, assuming $Q$, can give the answer. The advantage of the chart is that no calculation is involved.

### *Example*

A fabric-covered insulated surface has a heat loss $Q$ of 150 Btu/ft²-hr. The ambient temper-

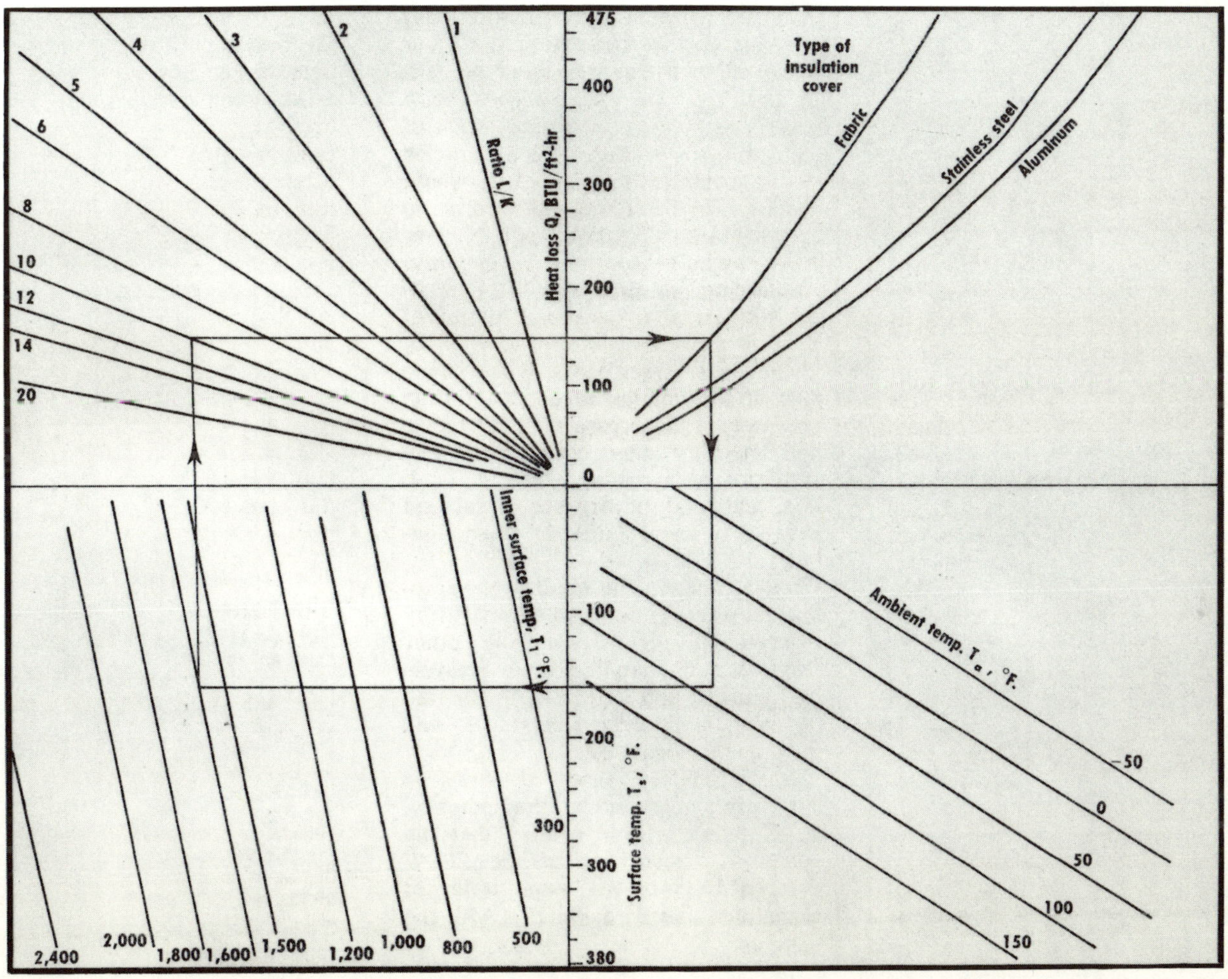

ature $T_a$ is 80°F and the inner surface temperature $T_1$ is at 1,600°F.

Determine the following:

1. Temperature at the surface $T_s$.
2. Thickness of insulation $L$, in.

***Solution***

On the chart, start at $Q = 150$ and move right to intersect the curve for fabric insulation cover. Then move down vertically to ambient temperature $T_a = 80$. Next move horizontally left to intersect the curve $T_1 = 1{,}600$. Read $T_s = 160$°F from the $T_s$ scale. Then move vertically to intersect a horizontal line through $Q = 150$ and read the ratio $L/K = 10$. If the insulation thermal conductivity $K$ is known, insulation thickness $L$ can then be easily obtained, as $L = 10/K$.

# 18. Charts estimate insulation thickness for preventing surface condensation

WHEN THE WATER VAPOR IN AIR COMES IN contact with surfaces that are at a temperature below the dew point of air, condensation occurs. This phenomenon must be considered when insulation thickness for flat surfaces or pipes is determined, to prevent condensation problems. The equation for steady-state heat transfer is

$$Q = \frac{t_a - t_d}{R_s}$$
$$= \frac{t_d - t_o}{L/k}$$

or, stated in terms of $L$,

$$L = R_s k \left(\frac{t_d - t_o}{t_a - t_d}\right) = R_s k Z$$

where:

$Q$ = heat loss, Btu/(sq ft)(hr)
$t_a$ = dry-bulb ambient temperature, F
$t_d$ = dew point, F
$R_s$ = surface thermal resistance, deg F(hr)(sq ft)/Btu
$t_o$ = operating temperature, F
$L$ = thickness of insulation for flat surfaces, in.

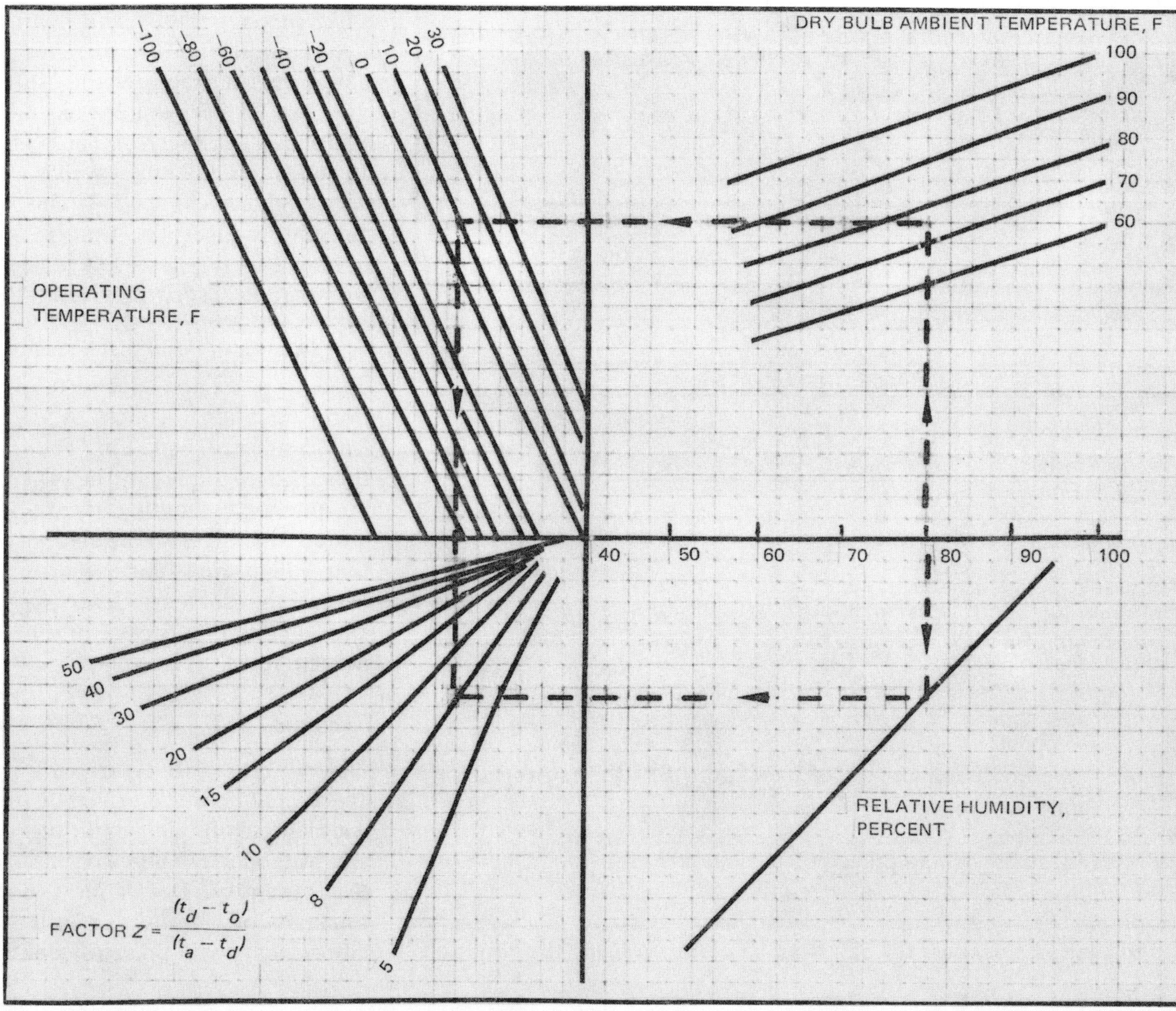

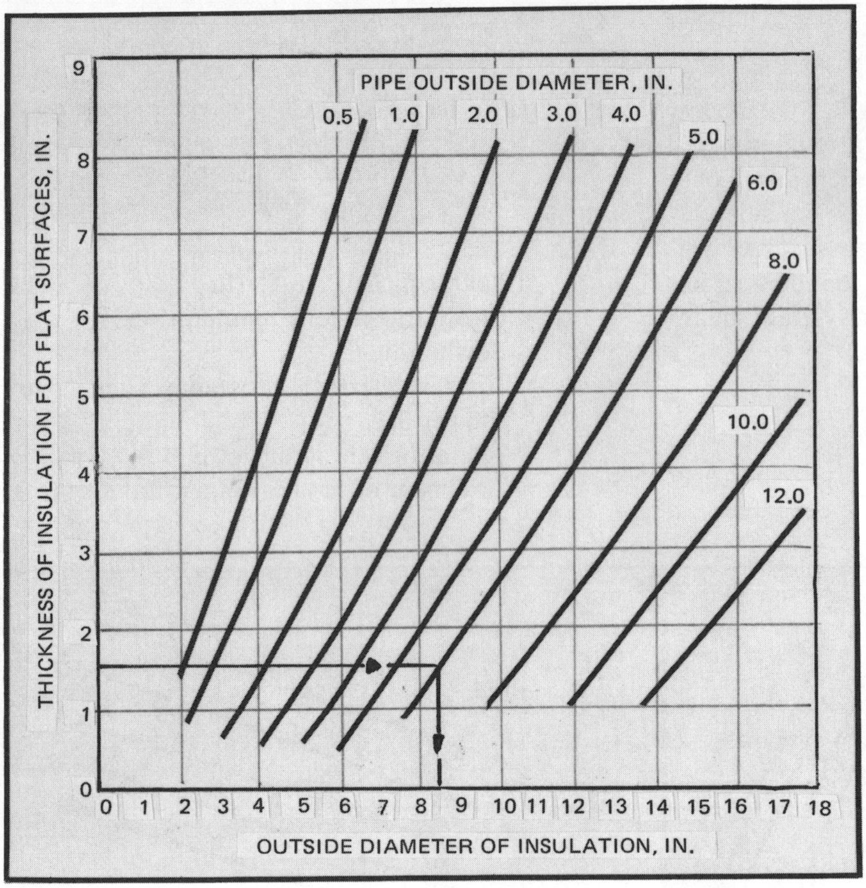


**SURFACE THERMAL RESISTANCE**

| Position of surface | Heat flow direction | Emissivity: Reflective e = 0.2 | Emissivity: Reflective e = 0.05 | Emissivity: Non-reflective e = 0.9 |
|---|---|---|---|---|
| Still air | | | | |
| Horizontal | Upward | 1.10 | 1.32 | 0.61 |
| Sloping (45 deg.) | Upward | 1.14 | 1.37 | 0.62 |
| Vertical | Horizontal | 1.35 | 1.70 | 0.68 |
| Sloping (45 deg.) | Downward | 1.67 | 2.22 | 0.76 |
| Horizontal | Downward | 2.70 | 4.55 | 0.92 |
| Moving air (any position) | | | | |
| 15 mph wind (winter) | | | | 0.17 |
| 7.5 mph wind (summer) | | | | 0.25 |

$k$ = thermal conductivity of insulation, Btu/(hr)(sq ft)(deg F/in.)

$$Z = \frac{t_d - t_o}{t_a - t_d}, \text{ dimensionless}$$

Insulation thickness for pipes can be obtained using the equation

$$L_p = \frac{d}{2} \ln \frac{d}{d_p}$$

where:

$L_p$ = pipe insulation thickness, in.
$d$ = outside diameter of insulation, in.
$= d_p + 2L_p$
$d_p$ = pipe outside diameter, in.

Although calculating thickness of insulation for flat surfaces appears to be simple, the real difficulty arises in estimating the dew point, after ambient temperature and relative humidity are known. This determination requires the use of steam tables, which can be avoided by using Nomograph 1. No reference to psychrometric charts is required. This method reduces the time required for problem solution. Nomograph 1 provides values of *Z*.

Nomograph 2 was developed to solve the pipe insulation equation, which is a transcendental equation. Thus these nomographs are helpful in quickly estimating the required thickness of insulation that will prevent surface condensation problems. Two examples illustrate the use of the nomographs.

***Example***

A horizontal flat surface is at a temperature of 10F. Ambient, still air conditions are 80F and 80 percent relative humidity. Determine the thickness of fibrous glass insulation that will prevent condensation problems when the thermal conductivity $k = 0.28$ Btu/(hr)(sq ft)(deg F/in.).

### *Solution*

On Nomograph 1, extend a vertical line down from 80 percent on Scale H to the reference line and up to the 80F line on Scale $t_a$. From that point of intersection, extend a line horizontally to the 10F line on Scale $t_o$ and then down vertically into Scale Z. Then, extend a horizontal line from the point on the reference line to intersect the previous line and read $Z = 9$. With $R_s = 0.61$ for a nonreflective horizontal surface (from the table), $L = R_s k Z = 0.61 \times 0.28 \times 9 = 1.54$ in. A safe solution is to use 2 in. of insulation.

### *Example*

All the conditions are the same as in the previous problem except that 6 in. dia pipe is used.

### *Solution*

On Nomograph 2, connect 1.54 in. on Scale $L$ with the 6 in. line on Scale $d_p$; from the point of intersection, extend a line to intersect Scale $d$, and read outside diameter of insulation as 8.6 in. Therefore, the thickness of insulation required for 6 in. dia pipe is $(8.6 - 6) \div 2 = 1.3$ in. Allowing a margin for safety, a 1.5 in. thickness should be used for insulating the pipe.

# 19. Chart estimates natural convection heat losses from surfaces

THIS NOMOGRAPH CAN BE USED TO ESTIMATE heat losses due to natural convection from surfaces. It is based on the equation: $Q = C \times 1/d^{0.2} \times 1/T^{0.181} \times \Delta T^{1.266}$, where $C$ = configuration-dependent constant (see key); $d$ = pipe or tube dia, in. ($d$ = 24 for a flat surface); $Q$ = heat loss, Btu/(h)(ft$^2$); $T$ = average of surface and ambient temperatures, $R$; and $\Delta T$ = difference between surface and ambient temperatures, $R$.

To use the figure, first connect $\Delta T$ with the appropriate code for the constant $C$, and extend to line Z. Then connect that point with the correct value for $d$, and extend to the $Q$ scale to read the answer. (This procedure can be reversed to find $\Delta T$ if $Q$ is given.)

### *Example*

Estimate the heat loss from a vertical flat surface at 600 R, when the air is at 500 R.

### *Solution*

Draw a line through $\Delta T$ = 100 and Point D on the C scale, and extend to Line Z. Connect that point with $d$ = 24 and extend to the $Q$ scale. $Q$ = 80 Btu/(h)(ft$^2$).

### *Reference*


Turner, W. C., and Malloy, J. F., "Thermal Insulation Handbook," published jointly by Robert E. Kriefer Publishing Co., Malabar, Fla., and McGraw-Hill Book Co., New York, 1981.


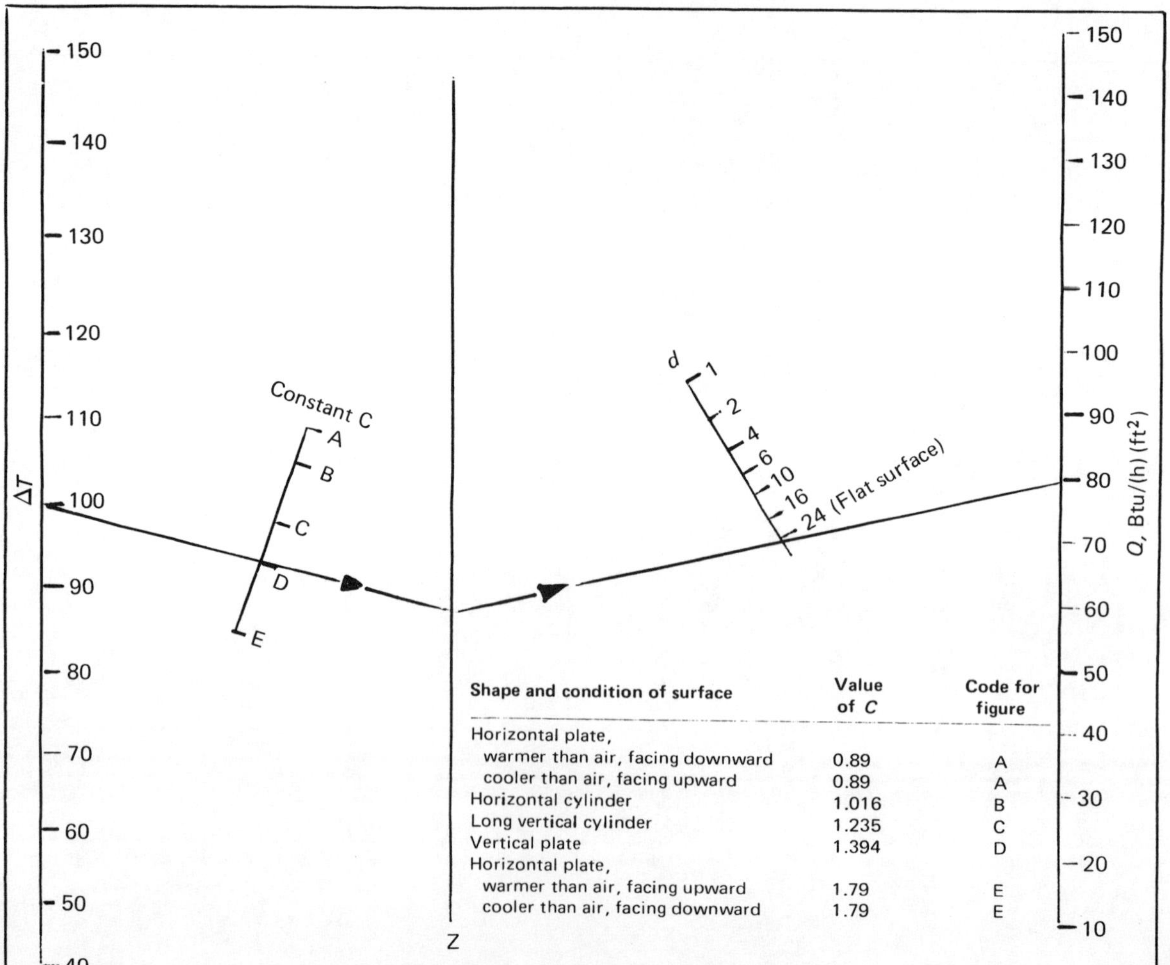


| Shape and condition of surface | Value of C | Code for figure |
|---|---|---|
| Horizontal plate, | | |
| warmer than air, facing downward | 0.89 | A |
| cooler than air, facing upward | 0.89 | A |
| Horizontal cylinder | 1.016 | B |
| Long vertical cylinder | 1.235 | C |
| Vertical plate | 1.394 | D |
| Horizontal plate, | | |
| warmer than air, facing upward | 1.79 | E |
| cooler than air, facing downward | 1.79 | E |

Estimate convective heat losses from surfaces

# 20. Chart evaluates performance of waste-heat boilers

PLANT ENGINEERS SHOULD KNOW WHAT PERformance to expect from waste-heat boilers at various inlet conditions, in order to predict boiler duty and check for fouling. The nomograph (Fig. 2) predicts the temperature of the gas leaving the boiler for any inlet condition and flowrate, based on the known performance of the boiler at other conditions. No heat-transfer calculations are needed, and the method applies to both firetube and watertube boilers (Fig. 1).

### *The Theory*

The duty ($Q$, kJ/h) of any boiler is:

$$Q = Wc_p(T_1 - T_2) = UA\Delta T \qquad (1)$$

where $W$ is gas flowrate (kg/h), $c_p$ is gas heat capacity (kJ/kg°C), $T_1$ and $T_2$ are the entering and exiting temperatures of the gas (°C), $U$ is the overall heat-transfer coefficient (kJ/m$^2$-h-°C), $A$ is the heat-transfer area (m$^2$), and $\Delta T$, the log-mean temperature difference, is given by:

$$\Delta T = \frac{T_1 - T_2}{\ln\left(\dfrac{T_1 - t_s}{T_2 - t_s}\right)} \qquad (2)$$

where $t_s$ is the saturation temperature of the steam at the boiler pressure. Substituting Equation 2 into Equation 1 and simplifying, we get a new equation relating $T_1$ and $T_2$:

$$\ln\frac{T_1 - t_s}{T_2 - t_s} = \frac{UA}{Wc_p} \qquad (3)$$

The overall heat-transfer coefficient is proportional to a power of the flowrate,* assuming that the temperature effect from one condition to another is minor:

$$U \propto W^{0.8} \text{ for a firetube boiler} \qquad (4)$$

$$U \propto W^{0.6} \text{ for a watertube boiler} \qquad (5)$$

Using Equation 4 later may cause some error, as tubewall temperature may change with flowrate, but Equation 5 will be accurate because in the watertube boiler the gas-film temperature will always be close to the steam temperature. Specific-heat variations can be ignored.

Combining Equations 3–5, we get equations for predicting boiler performance:

$$\ln\frac{T_1 - t_s}{T_2 - t_s} = K_1 W^{-0.2} \text{ for a firetube boiler} \qquad (6)$$

$$= K_2 W^{-0.4} \text{ for a watertube boiler} \qquad (7)$$

where $K_1$ and $K_2$ are constants for particular boilers in particular fouling states, assuming $c_p$ does not vary with temperature. Given one set of operating conditions and results, Equations 6 and 7 can be used to predict temperature or needed flowrate for another set of operating conditions.

Fig. 2 solves Equations 6 or 7 graphically. Since the $K$ values themselves are not important, they are indicated merely as points on the reference lines.

### *Example*

A watertube waste-heat boiler generates steam at a pressure of 980 kPa (142 psi). When gas

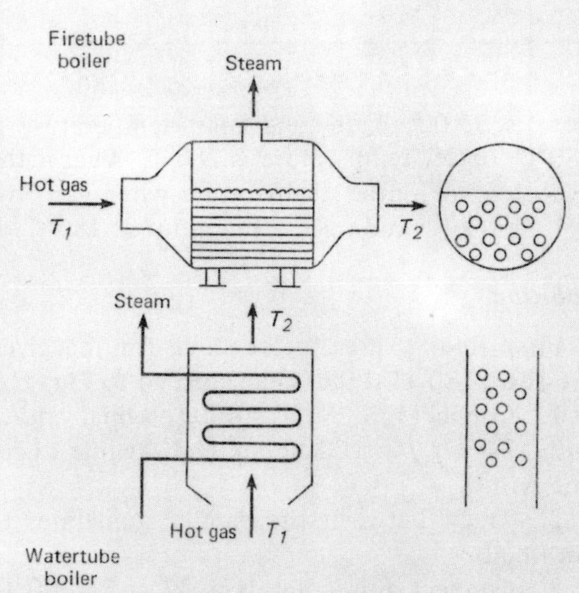


Fig. 1

*Ganapathy, V., Quick estimation of gas heat-transfer coefficients, *Chem. Eng.*, Sept. 13, 1976, p. 199.

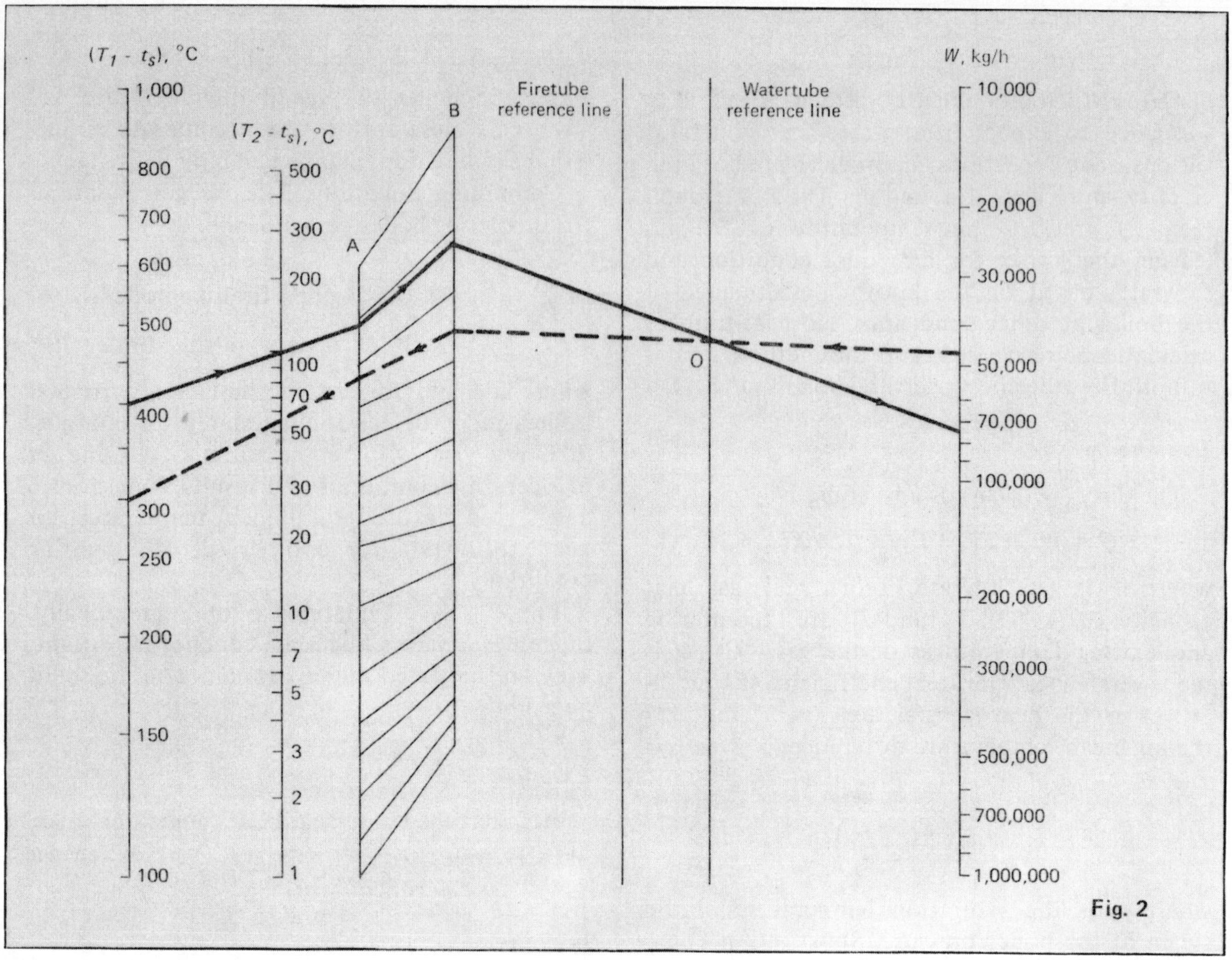

Fig. 2

flow is 75,000 kg/h, and inlet temperature is 580°C, outlet temperature is 280°C. What is the expected gas outlet temperature when gas flow is 45,000 kg/h and inlet temperature is 480°C?

***Solution***

First find $t_s$ in a steam table for the given pressure: 180°C at 980 kPa. Then go to Fig. 2:

1. Connect $(T_1 - t_s) = 400$ (original values) with $(T_2 - t_s) = 100$ and extend the line to cut line A.
2. Track between the parallel guidelines to cut line B.
3. Connect this point with $W = 75{,}000$ to find reference point 0 on the watertube-boiler line.

This reference point describes the initial operation of the boiler. To predict performance, we use the new conditions and backtrack:

1. Connect reference point 0 with $W =$ 45,000 (new value), extend to line B, and then to line A.
2. Connect this point to $(T_1 - t_s) = 300$, and extend to find $(T_2 - t_s) = 55$. Thus $T_2$, the new exit temperature, is 235°C if $U$ is what it was originally. If $T_2$ in actual operation is higher than 235°C, we must suspect that fouling has occurred.

# 21. Chart estimates fouling factor quickly

FOULING OF HEAT-TRANSFER SURFACES IS A phenomenon one has to live with. At best its effect can be reduced by generously designing the surfaces. Various factors such as velocity of heat-transfer media, dust concentration in flue gases, type (heavy or light oil) and nature of coal ash have an influence on fouling. A reduction in the duty of the equipment, and an increase in pumping power required, are results of fouling.

Plant engineers should have an idea of the magnitude of fouling factors for any heat-transfer equipment for the following reasons:

1. A shutdown for maintenance may be planned if fouling becomes serious.

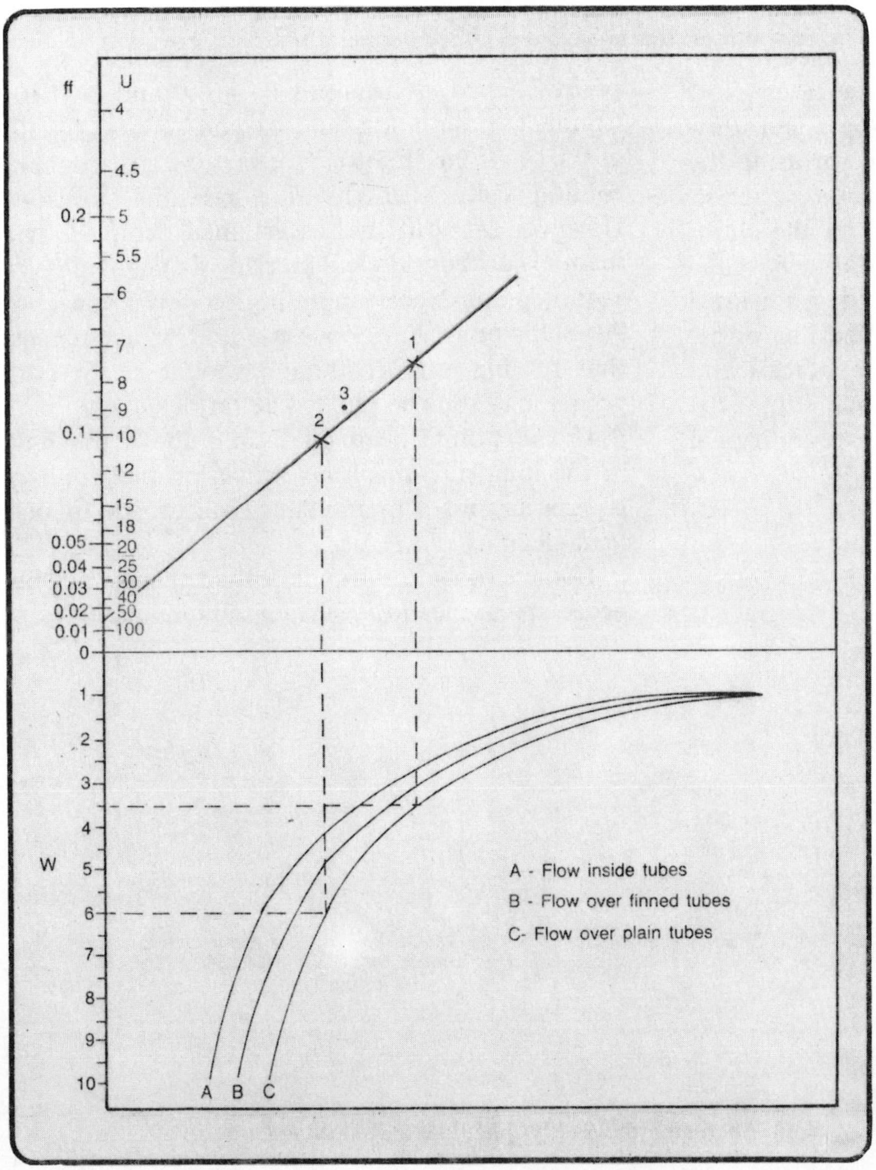


2. On-load cleaning methods may be used if fouling is only marginal, prolonging service until the planned shutdown period.

In any case it is helpful to know whether fouling is normal or severe. The accompanying chart can be used to check for fouling. Though approximate, it helps distinguish the trend.

### *Theory*

Assume that the overall heat-transfer coefficient is decided by one of the two media participating in heat transfer. This is commonly the case in economizers, waste-heat boilers, gas-liquid exchangers, and superheaters, where the air or gas side heat-transfer coefficient governs the overall heat-transfer coefficient.

The expresion for *U*, the overall heat-transfer coefficient, may be written as:

$$\frac{1}{U} = \frac{C}{W^n} + ff \qquad (1)$$

*C* is constant and *ff* is the total fouling factor. The index, *n*, depends on the type of flow. If the flow inside tubes governs *U*, then $n = 0.8$. If it is a water-tube boiler, superheater, or economizer with gas flow-

### *Nomenclature*

$C$ = Constant, function of gas properties

$ff$ = Fouling factor, ft$^2$-hr °F/Btu

$Q$ = Duty, Btu/hr

$S$ = Surface area, ft$^2$

$\Delta T$ = Log mean temperature difference, °F

$U$ = Overall heat transfer coefficient, Btu/ft$^2$-hr °F

$W$ = Flow of gas/air, lb/hr

ing over plain tubes, $n = 0.6$. For finned surfaces $n = 0.75$.

The advantage of estimating *ff* from the above is obvious. It is not necessary to know the gas properties nor solve complicated heat-transfer correlations. We may assume that the effect on heat transfer or change of fluid temperatures with load is marginal.

If $U$ is evaluated at two conditions of load, *ff* may be solved for.

$$U = \frac{Q}{S\Delta T} \tag{2}$$

$Q$ is the duty, $S$ the surface area and $\Delta T$, the log mean temperature difference. These values are known for a given duty.

The accompanying chart may be used to simplify the estimation of *ff* for three cases: flow inside tubes, flow over plain tubes, and flow over finned tubes. The procedure for using the chart is as follows:

Plot the values of $U$ and flow on the chart; connect them and extend to cut the *ff* scale. Read off the value. We may plot additional values and check if we are above the line or below. If above, the fouling is on the increase and something must be done. If below the line, the equipment may be operated for an additional time.

### *Example*

$U$ was found to be 7.5 Btu/ft$^2$-hr °F when 35,000 lb/hr of flue gases was flowing over the plain tubes in an economizer after a shutdown maintenance and cleaning. When 60,000 lb/hr was flowing, $U$ was 10 Btu/ft$^2$-hr °F.

Estimate the fouling factor.

After some months of operation it was found that with 50,000 lb/hr of flue gas, $U$ was 9 Btu/ft$^2$-hr °F. Is it a cause for concern?

### *Solution*

Any multiplication factor may be used for the flow scale. (Here, 10,000.) Go from $W = 3.5$ to cut curve C and project up until $U = 7.5$ is obtained. Mark this as point 1. Similarly mark point 2, corresponding to $W = 6.0$ and $U = 10$. Connect these points and extend to cut *ff* scale at 0.014 ft$^2$-hr °F/Btu. This may be a reasonable fouling factor for the flue gases in question. However, it can be ascertained from design manuals or from the engineering department.

Plot point 3 corresponding to $W = 5$ and $U = 9.0$. This point lies above the line, which means that fouling is more than it should be. Several points may thus be plotted at various loads.

If the points keep on falling above the line in the long run, this is an indication that fouling is serious and a maintenance shutdown or on-load cleaning is due.

If they lie below this line then it may not be necessary to clean the surfaces immediately.

# Chapter 4

# Valve Sizing, Piping and Equipment Design, Life Cycle Costing

# 1. Charts aid sizing of safety relief valves

SAFETY VALVES ARE REQUIRED BY THE AMERican Society of Mechanical Engineers' (ASME) codes for the protection of boilers and superheaters. It is common practice to size the relief valves on the superheater of drum-type boilers so that they relieve at least 20 percent of the maximum steaming capacity of the unit and allow the relief valves on the drum to release the remaining 80 percent.

However, the sum of the relieving capacity of all relief valves must always be greater than the maximum steaming capacity of the unit. And at least two valves are required on the drum and one on the superheater.

Plant engineers should be familiar with the relieving capacity of safety valves, or they should be able to quickly check the valve sizes required to relieve a particular quantity of steam.

The following formula may be used to estimate the relieving capacity of safety valves in the absence of specific relieving capacity information that is available from suppliers of equipment:

$$W = 51.5\, k_d\, P\, k_b\, k_s\, A \quad (1)$$

where:

$W$ = relieving capacity, lb/hr

$k_d$ = coefficient of discharge, approximately 0.96 for steam

$P$ = relieving pressure, psia

$k_b$ = correction factor for back pressure; unity for back pressures below 55 percent of absolute relieving pressure, which is generally the case (see Table 1 for representative values of $k_b$).

$k_s$ = correction factor for superheat, values given in Fig. 1; for saturated steam the value is unity.

$A$ = orifice area, sq in.; areas of standard orifices referred to by letter code in Table 2 are values supplied by manufacturers.

A simplified version of Equation 1 can be used by considering that $k_b$ and $k_d$ closely approach unity:

$$W = 49\, P A\, k_s \quad (2)$$

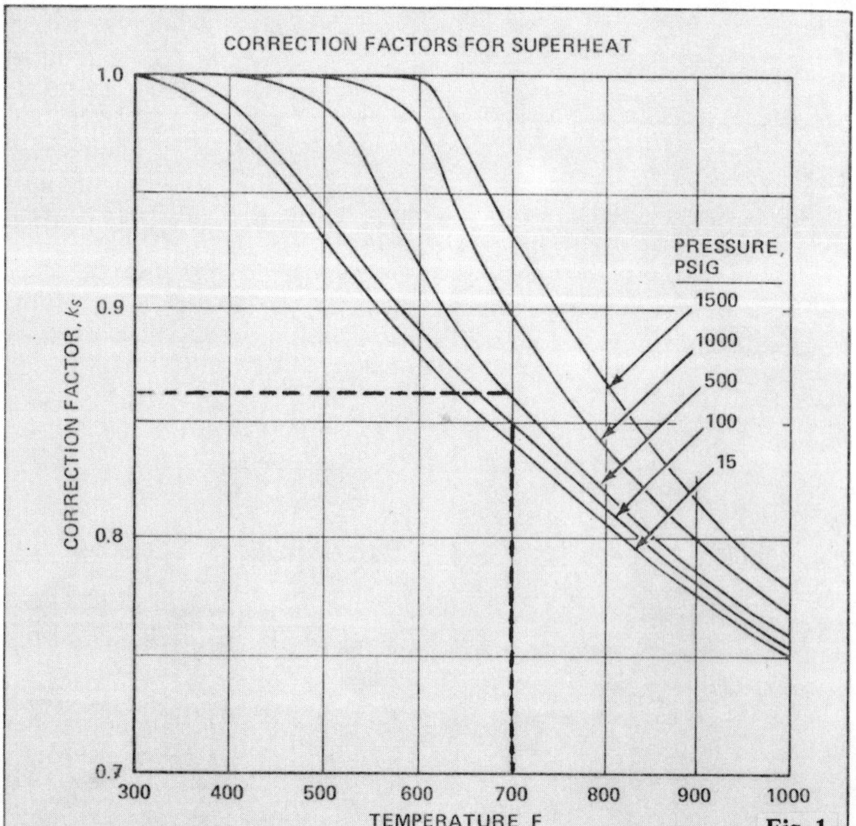


Fig. 1

Figure 2 is based on ASME Sec. I Power Boiler Procedure for 90 percent rating, so Equation 2 becomes

$$W = 44\, P A\, k_s \quad (3)$$

Section VIII, ASME Div. 1, sizes safety valves for 100 percent ratings, and Equation 2 conforms with Sec. VIII. Fig. 2

**Table 1. Ratio of Back Pressure to Set Pressure**

| *Back Pressure to Set Pressure Ratio* | *Back Pressure Correction Factor, $k_b$* |
|---|---|
| 0 to 0.55 | 1.0 |
| 0.6 | 1.0 |
| 0.7 | 0.94 |
| 0.8 | 0.84 |
| 0.9 | 0.64 |
| 1.0 | 0.00 |

can be used to solve Equation 3 with $k_s = 1.0$ (saturated steam). The advantage is that the orifice diameter in inches as well as the area in square inches is provided. This equation is helpful as it gives an approximation of the standard size that should be selected. If the required orifice diameter is 0.9 in., a 1.0 in. dia standard orifice (H orifice) may be used.

Two examples can be used to illustrate use of the charts.

**Table 2. Area of Standard Orifices Identified by Letter Code**

| *Orifice Symbol* | *Area, sq in.* |
|---|---|
| F | 0.297 |
| G | 0.503 |
| H | 0.785 |
| J | 1.29 |
| K | 1.84 |
| L | 2.85 |
| M | 3.6 |
| N | 4.34 |
| P | 6.38 |

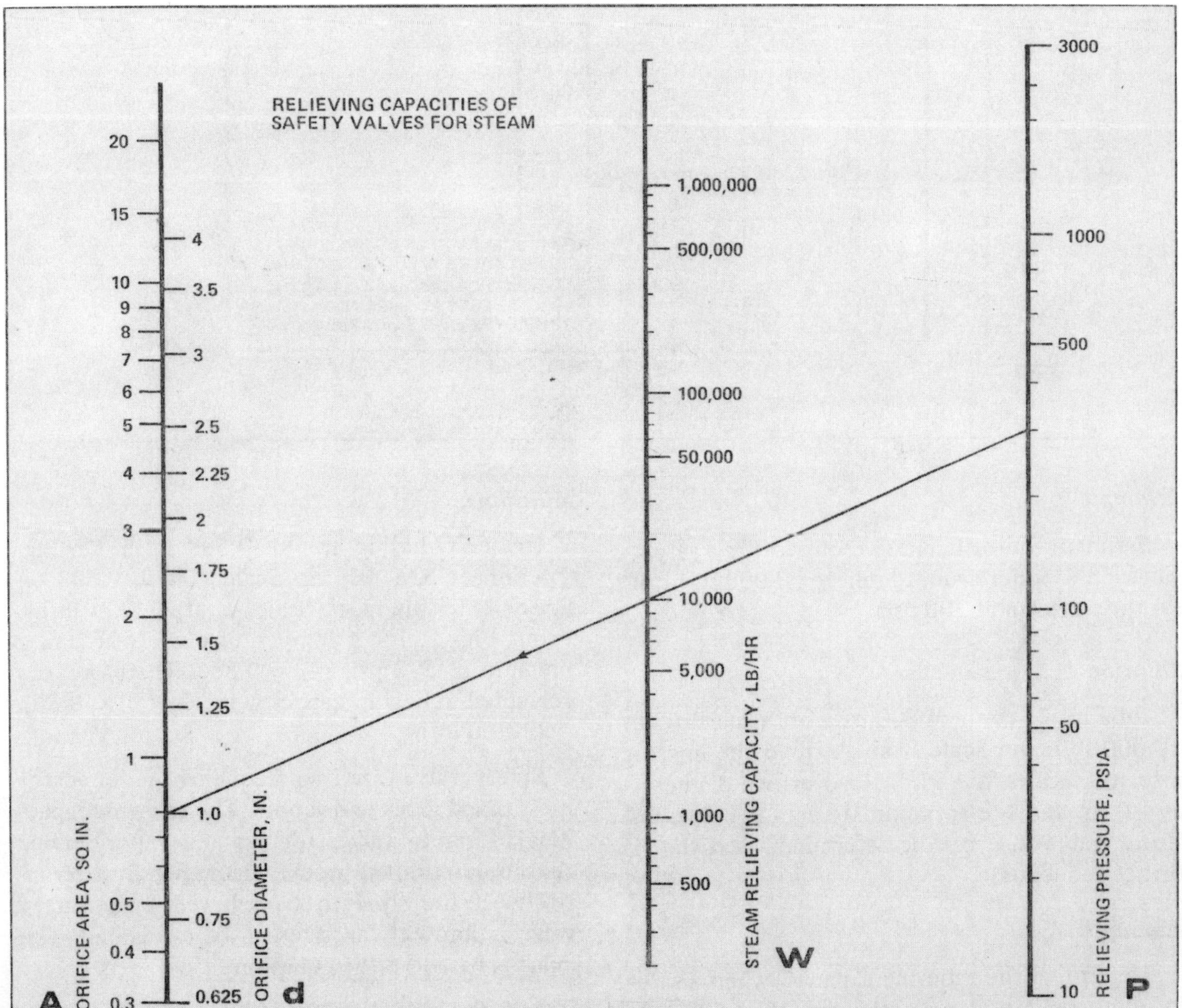


Fig. 2

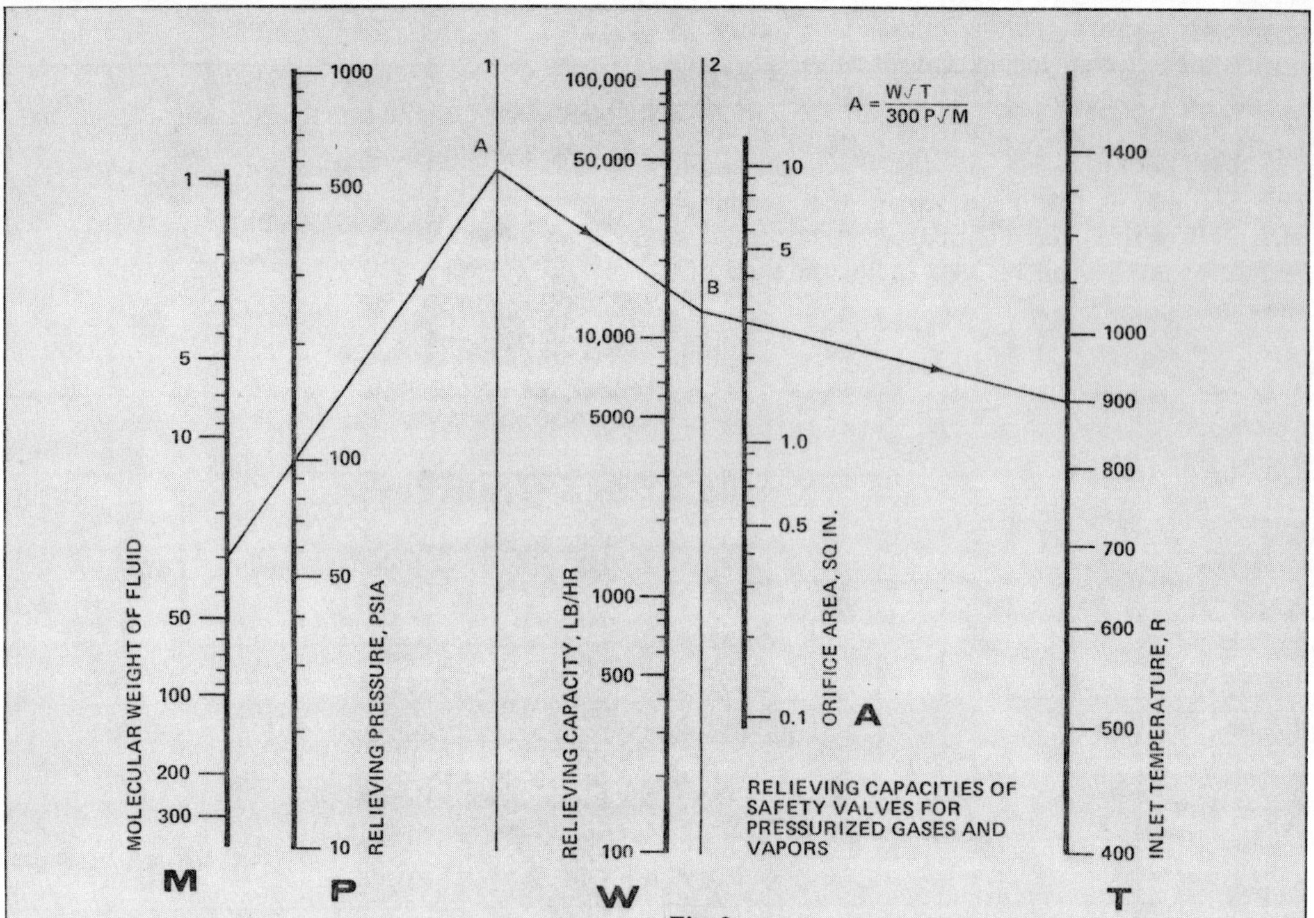

**Fig. 3**

***Example***

Estimate the orifice size required for a safety valve on a boiler drum to relieve 10,000 lb/hr of saturated steam at 300 psia.

***Solution***

On Fig. 2, connect 300 psia on Scale P with 10,000 lb/hr on Scale W and extend the line to intersect Scales A and d. Read orifice diameter as 1.0 in. and the approximate area as 0.8 sq in. From Table 2, it can be determined that the H orifice is adequate.

***Example***

Determine the relieving capacity when a safety valve on a boiler superheater set at 500 psig opens when the system is operated at 700F. Orifice used is shown as G in Table 2.

***Solution***

Orifice G has an area of 0.503 sq in. On Fig. 2, connect 515 psia on Scale P with 0.503 on Scale A to intersect Scale W at 13,000 lb/hr. From Fig. 1, for a temperature of 700F and a pressure of 500 psig, $k_s = 0.863$. Therefore, the corrected relieving capacity is 13,000 X 0.863 = 11,200 lb/hr.

Safety valves are also frequently required for pressurized gases and vapors. The relieving capacities of safety valves for gas and vapor service can be determined quickly using Fig. 3. Alternatively, if the flow to be relieved by a safety valve is known, the size can be calculated. The chart is based on the equation:

$$A = \frac{315\, W\sqrt{T}}{13.2\, P\sqrt{M}\sqrt{520\, C}} \qquad (4)$$

where:

$T$ = inlet temperature, R
$M$ = molecular weight of fluid
$C$ = constant, dependent on ratio of specific heats

Equation 4 may be simplified and rewritten, using a conservative value of 315 for $C$, as:

$$A = \frac{W\sqrt{T}}{300\,P\sqrt{M}} \qquad (5)$$

Fig. 3 is based on Equation 5.

### *Example*

Size a safety valve to relieve 15,000 lb/hr of air at 100 psia and 900R.

### *Solution*

On Fig. 3, connect 29 on Scale M with 100 psia on Scale P and extend the line to intersect reference line 1 at point A. Connect 15,000 lb/hr on Scale W with point A and extend the line to intersect reference line 2 at point B. Connect point B with 900R on Scale T, intersecting Scale A at 2.7 sq in. A valve with an orifice area of 2.7 sq in. will be adequate. From Table 2, the orifice symbol is L. The chart may be used in reverse to determine $W$ if $A$, $P$, $T$, and $M$ are known.

# 2. Chart provides control valve coefficients for steam

THIS NOMOGRAPH PROVIDES A SIMPLE MEANS for determining control valve coefficient $C_v$ when saturated and superheated steam flow conditions exist. Reference to steam tables is not necessary to determine the amount of superheat. Critical and noncritical flows have been considered so that solutions can be found on a single chart. The chart is based on these equations:*

For critical flow $(P_1 > 2P_2)$

$$C_v = \frac{W(1 + 0.00065\Delta T)}{1.85P_1} \qquad (1)$$

For noncritical flow $(P_1 < 2P_2)$

$$C_v = \frac{W(1 + 0.00065\Delta T)}{2.11\sqrt{\Delta P(P_t)}} \qquad (2)$$

where:

$C_v$ = control valve coefficient
$W$ = steam flow, lb/hr
$\Delta T$ = amount of superheat $(T - T_s)$, deg F
= 0 for saturated steam
$T$ = temperature of superheated steam, F
$T_s$ = temperature of saturated steam, F
$P_1$ = upstream pressure, psia
$P_2$ = downstream pressure, psia
$\Delta P$ = pressure drop $(P_1 - P_2)$, psia
$P_t$ = $P_1 + P_2$

### *Example*

In a case of noncritical flow where $P_1 = 1500$ psia, $P_2 = 1470$ psia, $\Delta P = 30$ psia, $W = 60{,}000$ lb/hr, and $T = 900$F, determine the valve coefficient $C_v$ for superheated steam.

### *Solution*

Align 30 psia on Scale $\Delta$P with 2970 on Scale $P_t$, intersecting Scale $P_1$ at 350 psia. Align 350 psia on Scale $P_1$ with 60,000 lb/hr on Scale W and extend a line to intersect reference line 1 at point A. Then, align 1500 psia on Scale $P_2$ with 900F on Scale T and extend the line to intersect reference line 3. Follow the guidelines to reference line 2 at point B. Align points A and B and, where Scale $C_v$ is crossed, read valve coefficient for superheated steam as 110. For saturated steam, align point A with point C on reference line 2 and, where Scale $C_v$ is crossed, read valve coefficient as 88.

### *Example*

For critical flow, where $P_1 = 200$ psia, $P_2 = 90$ psia, $W = 20{,}000$ lb/hr, and $T = 500$F, find $C_v$ for superheated steam.

### *Solution*

Align 200 psia on Scale $P_1$ with 20,000 lb/hr on Scale W and extend a line to intersect reference line 1 at point D. Then, align 90 psia on Scale $P_2$ with 500F on Scale T and extend the line to intersect reference line 3. Follow the guidelines to reference line 2 at point E. Align points E and D and, where Scale $C_v$ is crossed, read valve coefficient for superheated steam as 57.

*Copes-Vulcan, Bulletin 1122, "Direct Steam Converting Valve."

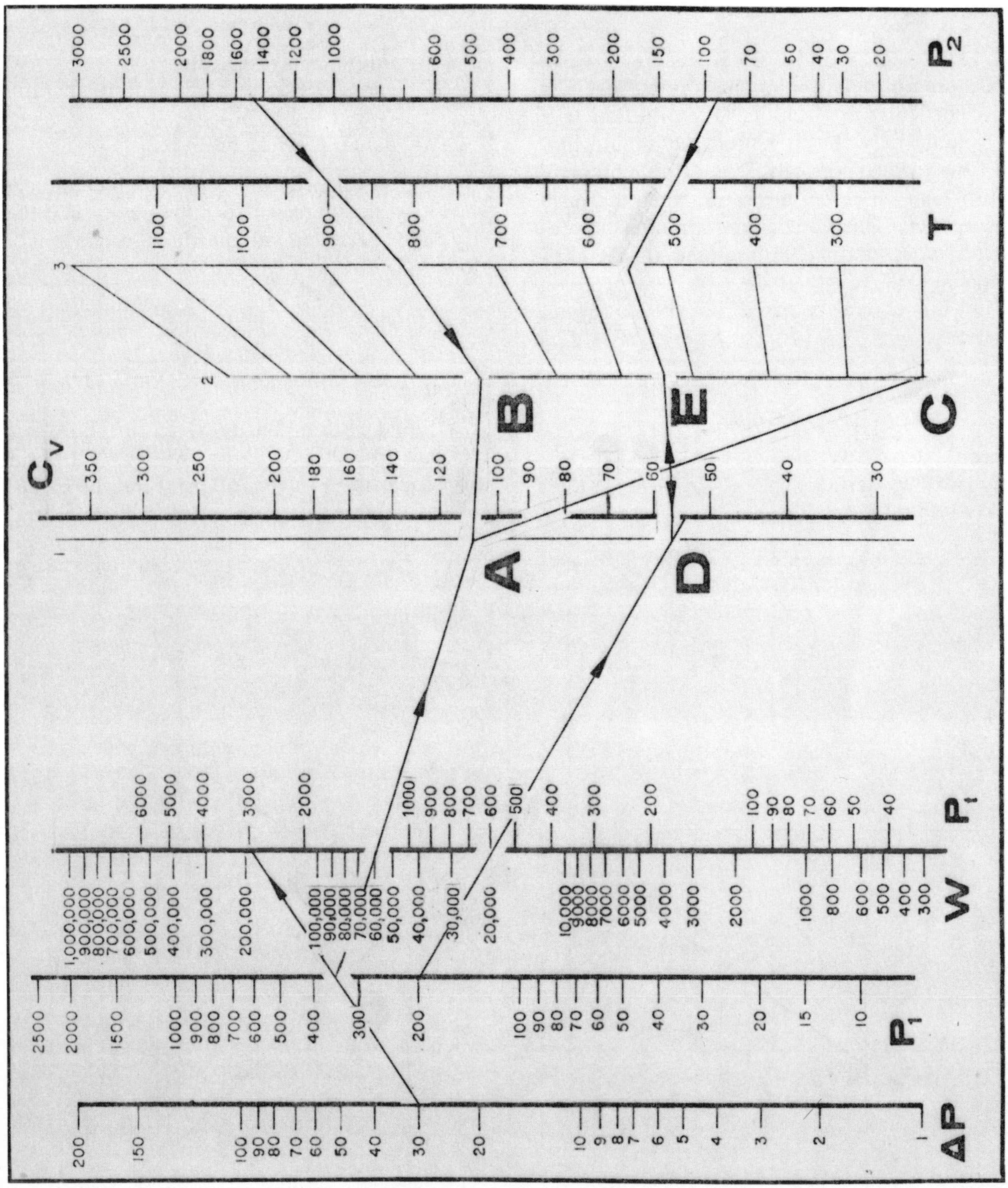
ΔP
P1
W
P1
A
B
C
D
E
Cv
T
P2

# 3. Nomogram estimates control valve coefficient for water, liquids

THE FLOW COEFFICIENT OF A VALVE IS COMmonly used in determining the correct valve size for a particular application. By definition it is the volume of water in gallons per minute that will flow through the valve body with a pressure drop of one psi across the valve, when the valve is wide open. This nomogram simplifies the procedure for estimating the required flow coefficient.

The nomogram is based on the following equations:

$$C_v = V(G/\Delta P)^{1/2}$$

$$V = W/500G$$

Where:

$C_v$ = flow coefficient
$V$ = liquid flow rate, gpm
$W$ = liquid flow rate, lb/hr
$G$ = specific gravity at flowing temperature (water = 1 at 60F)
$\Delta P$ = actual pressure drop, psi

For convenience, the nomogram includes conversion scales between specific gravity and density and between volumetric and gravimetric flow.

### *Example*

A liquid with a density of 45 lb/ft$^3$ flows at the rate of 100,000 lb/hr. Allowable pressure drop for sizing a valve is 50 psi. Find the specific gravity, volume flow rate and flow coefficient.

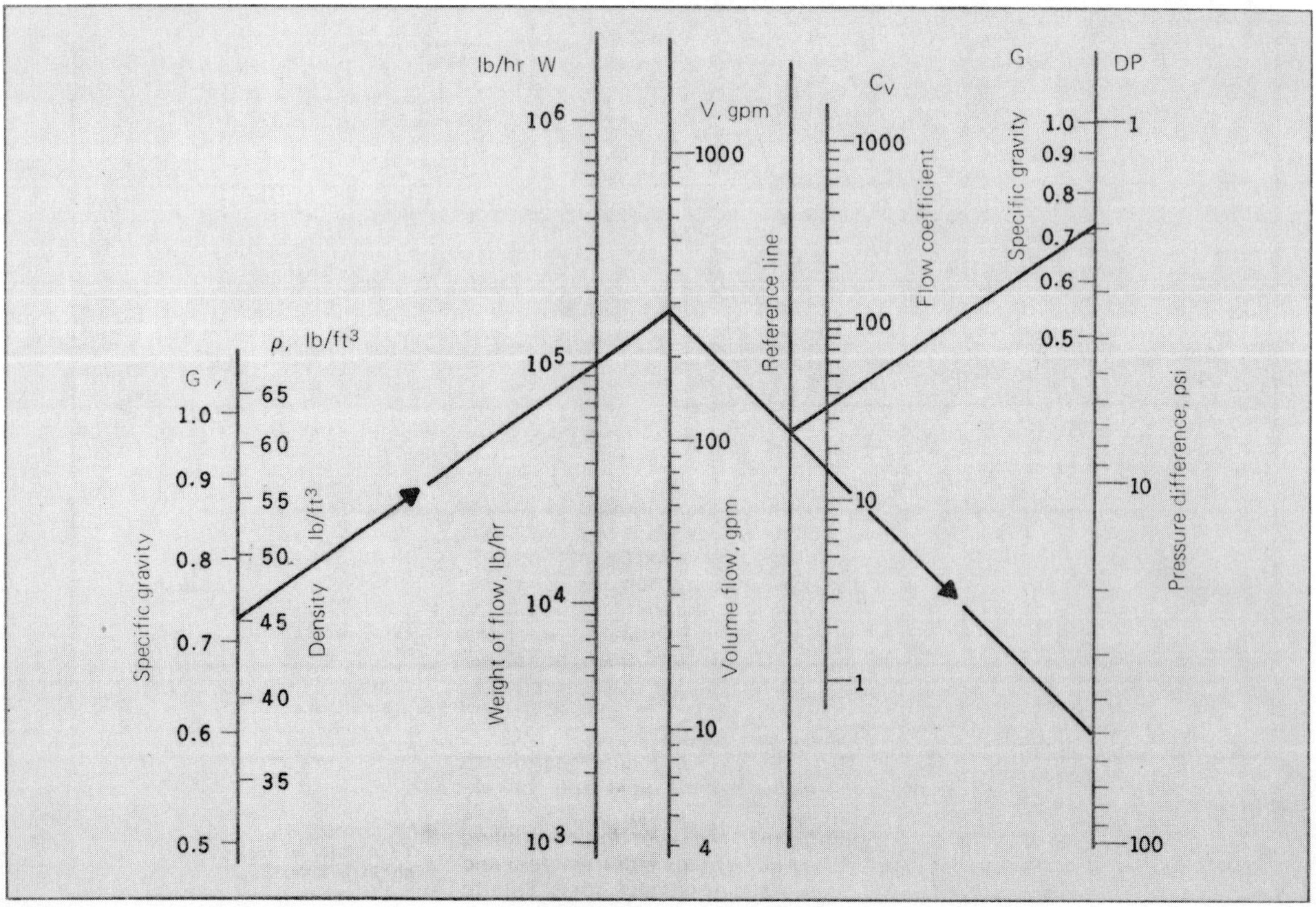

***Solution***

From the left-hand scale, specific gravity is read off directly as 0.72. This point is connected with the flow rate of 100,000 lb/hr on the W scale and the line extended to the V scale to obtain 280 gpm. This point on the V scale is connected with 50 on the pressure-drop scale. The intersection of this line with the reference line is connected with the specific gravity value on the right-hand scale and the intersection with the flow coefficient scale gives 34 as the answer.

If volume flow is given rather than weight flow, the solution remains the same except that volume flow is used to draw the first line.

# 4. Charts for determining flowmeter sizes

PLANT ENGINEERS OFTEN ARE REQUIRED TO measure liquid or gas volumes flowing through pipes and ducts. Numerous flowmeters are available for this purpose, but complicated calculations are necessary to determine the size of the instrument needed for a particular application.

The basic equation for calculating volume through a flowmeter is:

$$W = \frac{359YCd^2}{\sqrt{1 - B^4}} \sqrt{vh}$$

where:

$W$ = flow volume, lb/hr
$Y$ = liquid or gas expansion factor
$C$ = discharge coefficient
$d$ = flowmeter diameter, in.
$B$ = $d/D$ ratio
$D$ = pipe (or duct) diameter, in.
$v$ = liquid or gas density, lb/cu ft
$h$ = pressure drop across flowmeter, in. w.g.

The equation is difficult to use if $d$ and $B$ are unknown. When $W$, $C$, $v$, and $h$ are known, trial and error calculations can be performed to solve for $d$.

The accompanying nomographs, developed from the rearranged equation

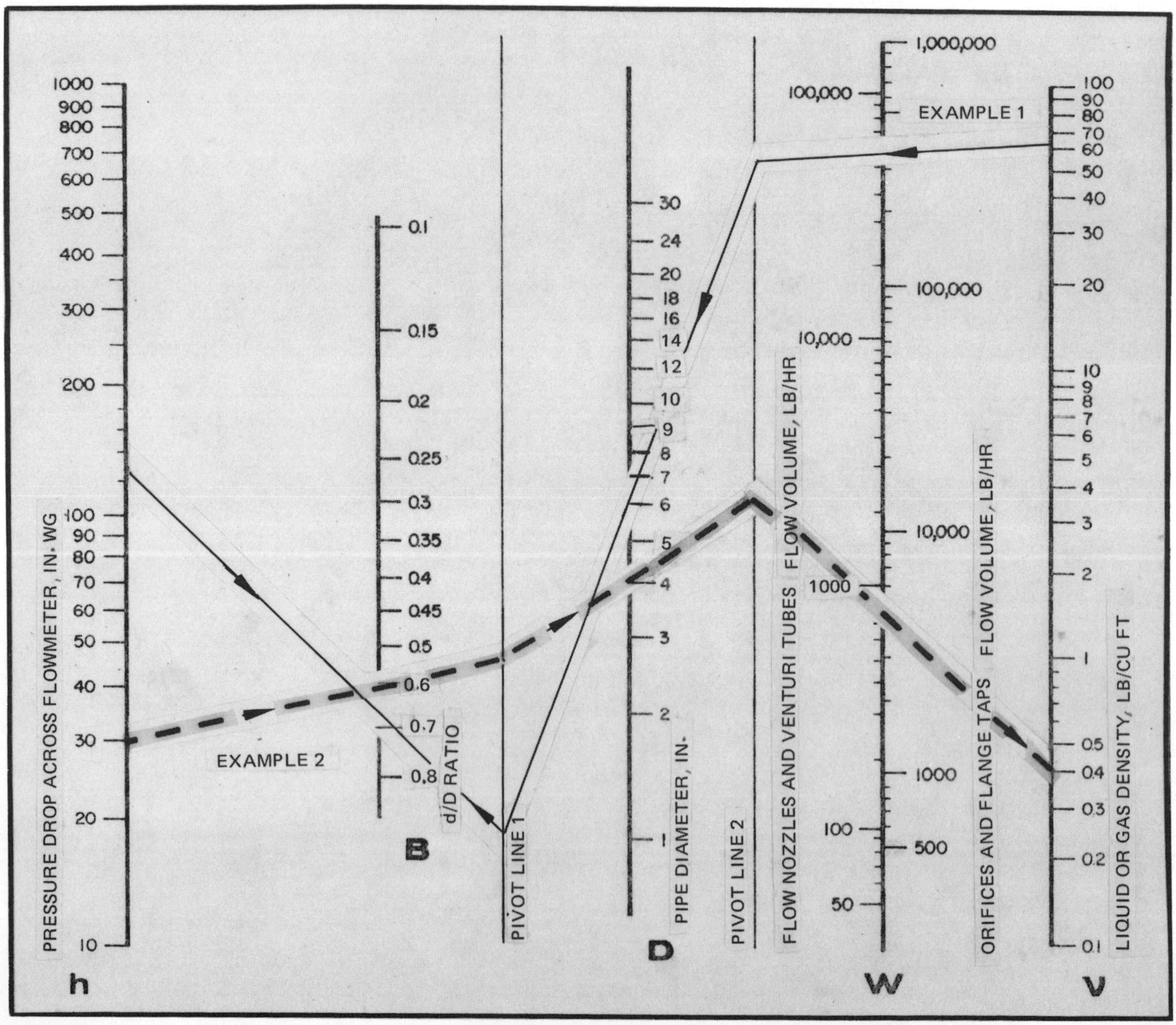


**Nomograph 1. Gas expansion factors are dependent on pressure and dimensional data.**

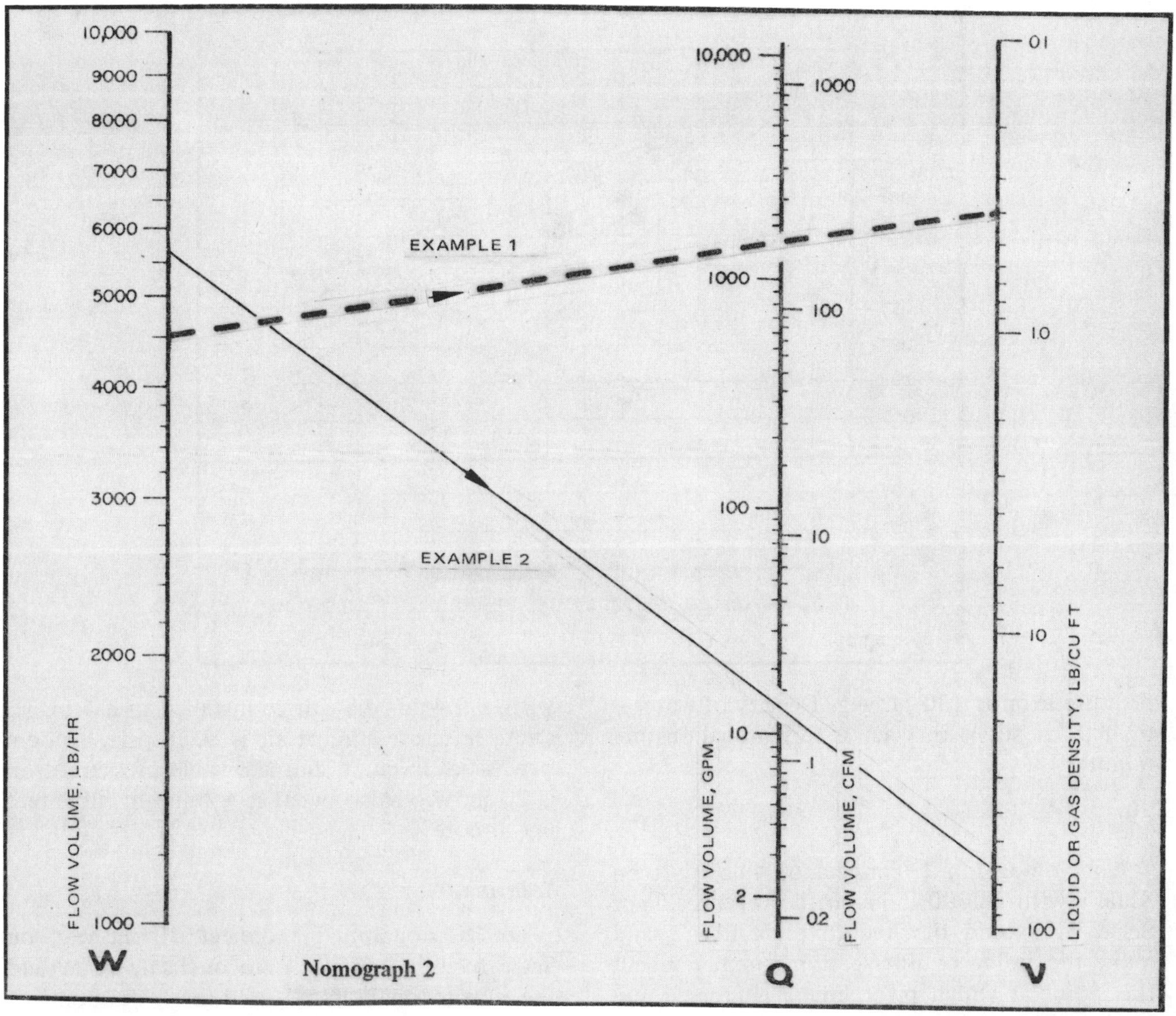


**Nomograph 2**

$$W = 359CYD^2 \frac{B^2}{\sqrt{1 - B^4}} \sqrt{vh}$$

provide a convenient method for sizing flow nozzles, orifices, venturi tubes, and flange taps. Conversely, if flowmeter size is known, flow rate and pressure drop across the device can be determined.

The following assumptions were made in developing the nomographs:

- $C = 0.61$ for orifices and flange taps; 0.98 for flow nozzles and venturi tubes
- Effect of Reynolds number on $C$ is negligible (maximum error is 2 percent)
- $Y$ is dependent on flowmeter type and specific heat ratios. The expansion factor graph disregards flowmeter type and specific heat ratios (maximum error is 3 percent).

These symbols are used in addition to those appearing in the equation:

$Q$ = flow volume, gpm (liquids), cfm (gases)
$P$ = flowmeter inlet pressure, psia

### *Example 1*

Determine the size of a venturi tube needed to measure a flow of 58,000 lb/hr of 60F water through a 6 in. ID pipe. Maximum allowable

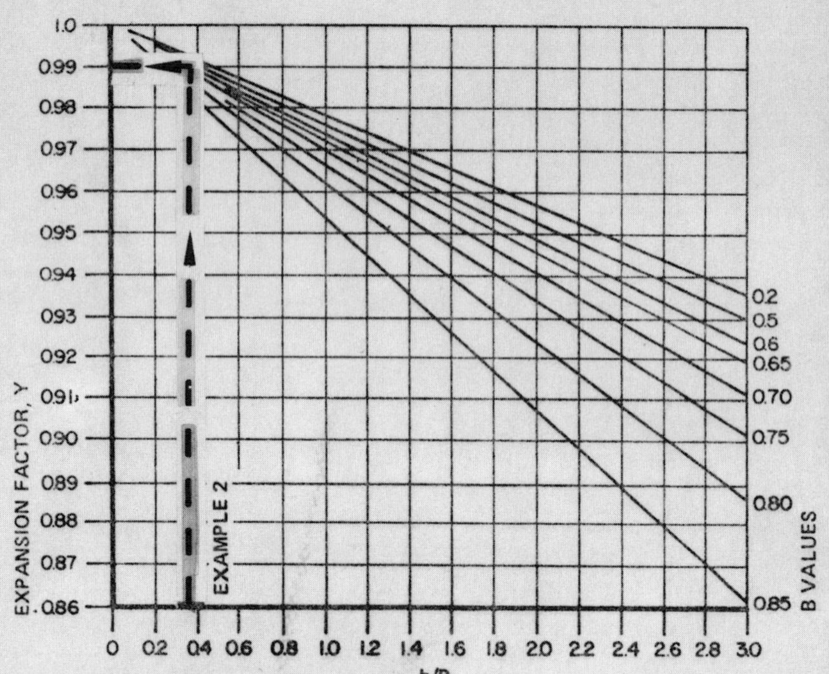


Nomograph 3

pressure drop is 130 in. w.g. Density of water is 62.4 lb/cu ft. Also, what is flow in gallons per minute?

***Solution***

On Nomograph 1, connect 62.4 lb/cu ft on Scale v with 58,000 lb/hr on the venturi side of Scale W, extend the line to pivot line 2, and mark the point of intersection. Next, connect this point at which pivot line 1 is crossed, extend a line to 130 in. on Scale h, and where Scale B is intersected, read $d/D$ ratio as 0.67. Hence, the venturi tube diameter $d$ is equal to 0.67×6.0, or 4.02 in.

Next, on Nomograph 2, connect 5800 lb/hr (58,000 reduced by a factor of 10) on Scale W with 62.4 lb/cu ft on Scale v, and, where the line intersects Scale Q, read 11.6 gpm. Scales W and Q incorporate the same multiplication factor; therefore, because 58,000 was reduced by a factor of 10, 11.6 must be multiplied by the same factor and 116 gpm of water is flowing through the venturi tube.

***Example 2***

Estimate the air flow that can be measured with a 2.4 in. dia orifice installed in a 4 in. ID duct. Inlet pressure of air is 80 lb psia, air density is 0.4 lb/cu ft, and allowable pressure drop is 20 in. w.g. Also, what is volume in cubic feet per minute?

***Solution***

On Nomograph 1, connect 30 in. w.g. on Scale h with 0.6 $d/D$ ratio on Scale B, extend the line to pivot line 1, and mark the point of intersection. Connect this point with 4 in. on Scale D and extend the line to pivot line 2. From the point at which pivot line 2 is crossed, extend a line to 0.4 lb/cu ft on Scale v, and, where Scale W is intersected, read air flow on orifice side as 4550 lb/hr.

But, this air flow value must be corrected by an expansion factor. Using $B = 0.6$ and $h/P = 0.375$ (30÷80), a corresponding $Y$ value of 0.99 is obtained from the expansion factor graph. Hence, the corrected air flow is 4550×0.99, or approximately 4500 lb/hr.

Returning to Nomograph 2, connect 4500 lb/hr on Scale W with 0.4 lb/cu ft on Scale v, and, where Scale Q is intersected, read air flow volume as 190 cfm.

# 5. Chart corrects steam-flow meter readings for different parameters

STEAM-FLOW METERS SUCH AS NOZZLES AND orifices are usually calibrated for specific steam conditions. Plants frequently operate at different steam conditions so corrections to the steam-flow readings must be made to obtain accurate data. Steam tables listing specific volumes are normally used to make such corrections.

The nomograph on the following page determines steam flow at any steam condition if the flow at a specific steam condition is known. The nomograph is based on the equation:

$$W = \frac{359CYd^2\sqrt{H/V}}{\sqrt{1-(d/D)^4}}$$

where:

$W$ = steam flow, lb/hr
$C$ = coefficient of discharge
$Y$ = expansion factor
$d$ = orifice diameter, in.
$H$ = differential pressure across the flow meter, in. w.g.
$V$ = steam specific volume, cu ft/lb
$D$ = pipe diameter, in.

For a given flow meter, $d$, $C$, and $D$ are constants. The effect of changes in $Y$ may be neglected because it is marginal. Thus the major factors contributing to steam flow are the heat (differential pressure) and specific volume. Simplifying the previous equation;

$$W\sqrt{V/H} = \text{a constant}$$

The nomograph easily solves this equation.

### *Example*

A steam-flow meter shows a reading of 9000 lb/hr when steam conditions are 1000 psia at 800F and the differential pressure is 300 in. w.g. Plant operating conditions require a change to 800 psia (saturated conditions) and 200 in. w.g. What is the new steam flow?

### *Solution*

Draw a vertical line from 800F on Scale T to the 1000 psia curve. Draw a horizontal line from the intersected point to Reference Line 1. Connect this point with 90,000 lb/hr on Scale W (a multiplication factor of 10 is used in Scale W) and extend the line to Reference Line 2. Connect the marked point with 300 in. w.g. on Scale H. Mark the point where the line crosses Reference Line 3.

Draw a line from 200 in. w.g. on Scale H to the intersection point on Reference Line 3 and extend this line to Reference Line 2. Mark the intersection point on Reference Line 2.

Extend a horizontal line from the 800 psia saturation line to Reference Line 1. Draw a line from the 200 in. w.g. line intersection point on Reference Line 2 to the intersection point on Reference Line 1 and read 80,000 lb/hr where the line crosses Scale W. Divide this value by 10 (the Scale's multiplication factor) to obtain the new steam flow, 8000 lb/hr.

*/ Chart corrects steam-flow meter readings for different parameters /*

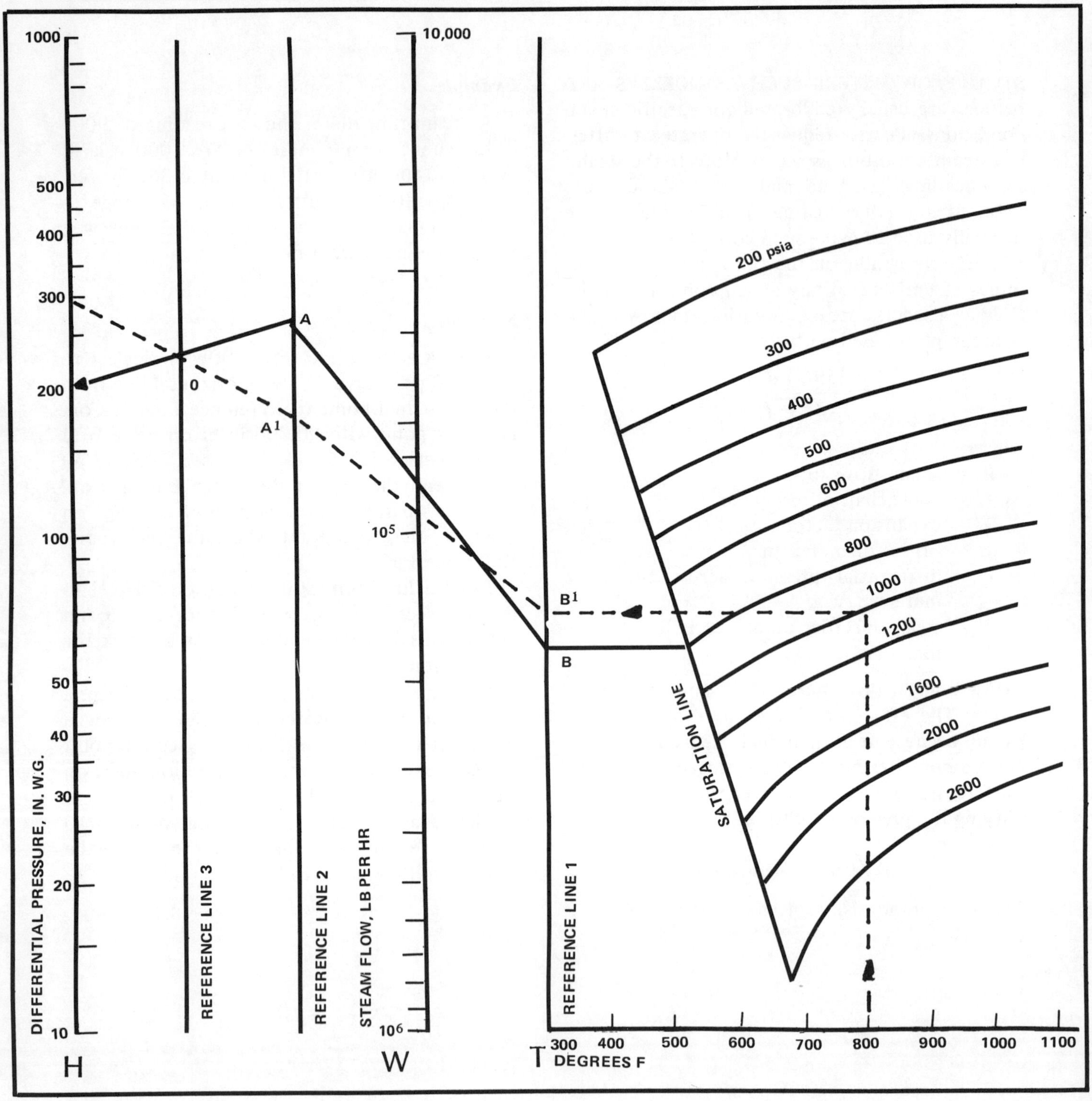

# 6. Nomogram provides quick estimate of pressure vessel thickness

QUICK ESTIMATES OF SHELL THICKNESS PER ASME Code Section VIII or, alternatively, the maximum pressures a given shell can withstand, may be made with this nomogram. It is based on the following equation (Ref. 1):

$$t = PR/(SE - 0.6P)$$

where:

$t$ = shell thickness, in.
$P$ = internal design pressure, psig
$R$ = inner radius of shell, in.
$S$ = allowable stress, psi
$E$ = ligament efficiency

Tables 1 and 2 (from Ref. 1) list the allowable stresses for commonly used carbon steels, and the ligament efficiencies for common methods of fabrication and inspection.

*Example*

A pressure vessel with a design pressure of 300 psig and an inner diameter of 28 in. is to be made of carbon steel, SA 516 grade 60, with double welded butt joints, spot examined. Estimate the minimum shell thickness.

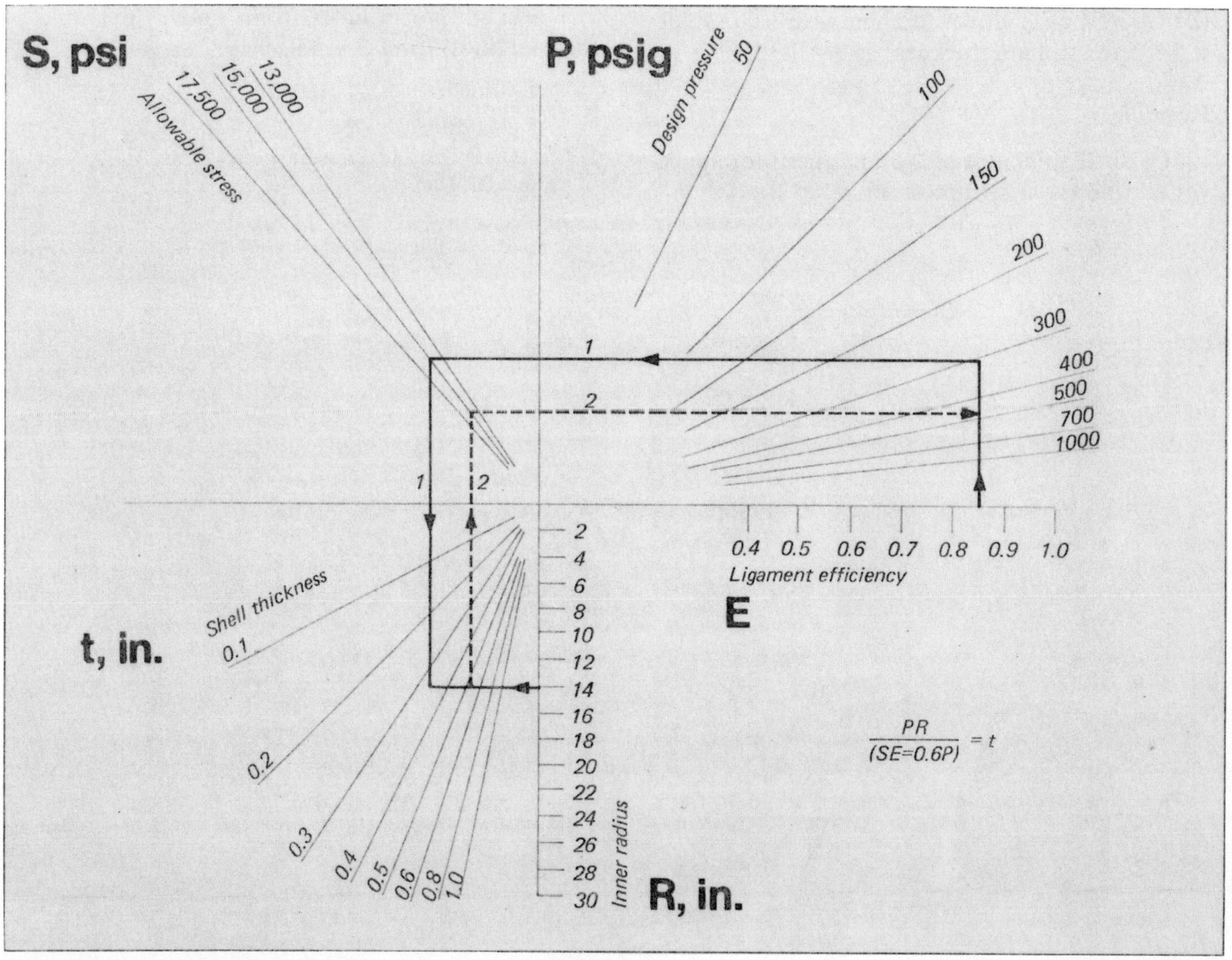

**Table 1. Allowable Stress**

| *Material* | *Grade* | *Allowable Stress (up to 650F) psi* |
|---|---|---|
| SA 515 or SA 516 | 55 | 13,800 |
| | 60 | 15,000 |
| | 65 | 16,300 |
| | 70 | 17,500 |
| SA 106 | B | 15,000 |
| | C | 17,500 |

**Table 2. Ligament Efficiency *(E)***

| | *Fully Radiographed* | *Spot Exam.* | *Not Spot Exam.* |
|---|---|---|---|
| Butt joints, double welded | 1.0 | 0.85 | 0.70 |
| Single welded butt joint, with backing strip | 0.9 | 0.8 | 0.65 |
| Single welded butt joint, without backing strip | – | – | 0.60 |

### *Solution*

From the tables $E = 0.85$ and $S = 15{,}000$. (1) Go vertically up from $E = 0.85$ to cut $P = 300$. (2) Go horizontally to cut $S = 15{,}000$. (3) Go vertically down to align with value of $R = 14$, and estimate thickness as $t = 0.33$.

### *Example*

If a shell thickness of 0.5 in. is used for above vessel, find the design pressure.

### *Solution*

Working in reverse from above example (dashed lines), find $P = 450$ psig.

### *Reference*

1. ASME Boiler and Pressure Vessel Code, 1980, Section VIII, Div. 1.

# 7. Graphical method for designing pressure vessels for external pressure

THE STANDARD PROCEDURE FOR DESIGNING pressure vessels and tubes (or pipes) subject to external pressure is tedious and time-consuming, because it involves a trial-and-error approach, and reference to many charts and tables. However, by using the method explained in this article, and Figs. 1 and 2, quick estimates of shell or tube thickness, or external design pressure, can be obtained for a wide variety of materials without the need for a complicated iterative procedure.

Many chemical process equipment items must be designed for external pressure. For example, process vessels often may be subjected to accidental vacuum. The ASME Boiler and Pressure Vessel Code requires that these be treated as vessels subject to external pressure. Vessels with jackets, depending on the pressure in the jackets, may have to be designed for external pressure.

In heat exchangers, evaporators, aftercoolers, etc., where there is high pressure on the shell side, the tubes must be designed for external pressure.

To show the advantage of this new method, let us first describe the procedure of the ASME Boiler and Pressure Vessel Code, Sec. VIII, Div. 1, Para. UG-28.

### *ASME Procedure*

*For cylinders, shells and tubes where* D/t > 10:

Step 1—Assume a value for $t$, and determine $L/D$ and $D/t$.

Step 2—Enter Fig. UGO-28 (in the ASME manual) at the value of $L/D$. For $L/D>50$, use $L/D=50$.

Step 3—Move horizontally to the line for the value of $D/t$, and from the point of intersection, read off $A$.

Step 4—Using $A$, enter the appropriate chart for material and temperature and read off $B$ (from Fig. UCS 28-1 or 28-2 in the Code).

Step 5—Calculate $P_a = 4B/[3(D/t)]$.

Step 6—If $P_a$ is less than the design pressure, $P$, increase $t$ and repeat Steps 1 to 6.

*For cylinders, shells and tubes where* D/t < 10:

Step 1—Using the same procedure as before, find $A$ and $B$. For $D/t$ less than 4, $A=1.1/(D/t)^2$. For $A>0.1$, use 0.1.

Step 2—Calculate $P_{a1}$ and $P_{a2}$.

$$P_{a1} = \left[\frac{2.167}{D/t} - 0.0833\right] B$$

$$P_{a2} = \frac{2S}{D/t}\left(1 - \frac{1}{D/t}\right)$$

where $S$ is the lesser of: twice the maximum allowable stress value at the design metal temperature (from the applicable table in Subsection C), or 0.9 times the tabulated yield strength of the material at the design temperature (values of yield strength are available in Section VIII, Div. 2, Table ACS 2).

### *Nomenclature*

| | |
|---|---|
| $A, B$ | Factors from charts in ASME Boiler and Pressure Vessel Code, Section VIII, Div. 1 and 2: Fig. UGO-28.0, UCS-28.1 and UCS-28.2 |
| $D$ | Outer dia of shell or tube, in. |
| $L$ | Design length of shell or tube, in. |
| $P$ | External design pressure, psi |
| $P_a$ | Maximum allowable external pressure on shells or tubes, psi |
| $S$ | Lesser of: twice the maximum allowable stress value at design temperature, or 0.9 times the tabulated yield strength at the design temperature. (Values of tabulated yield strengths in Section VIII, Div. 2 may be used in the absence of such values in Div. 1.) |
| $t$ | Minimum thickness of shell or tube, exclusive of corrosion allowance, in. |

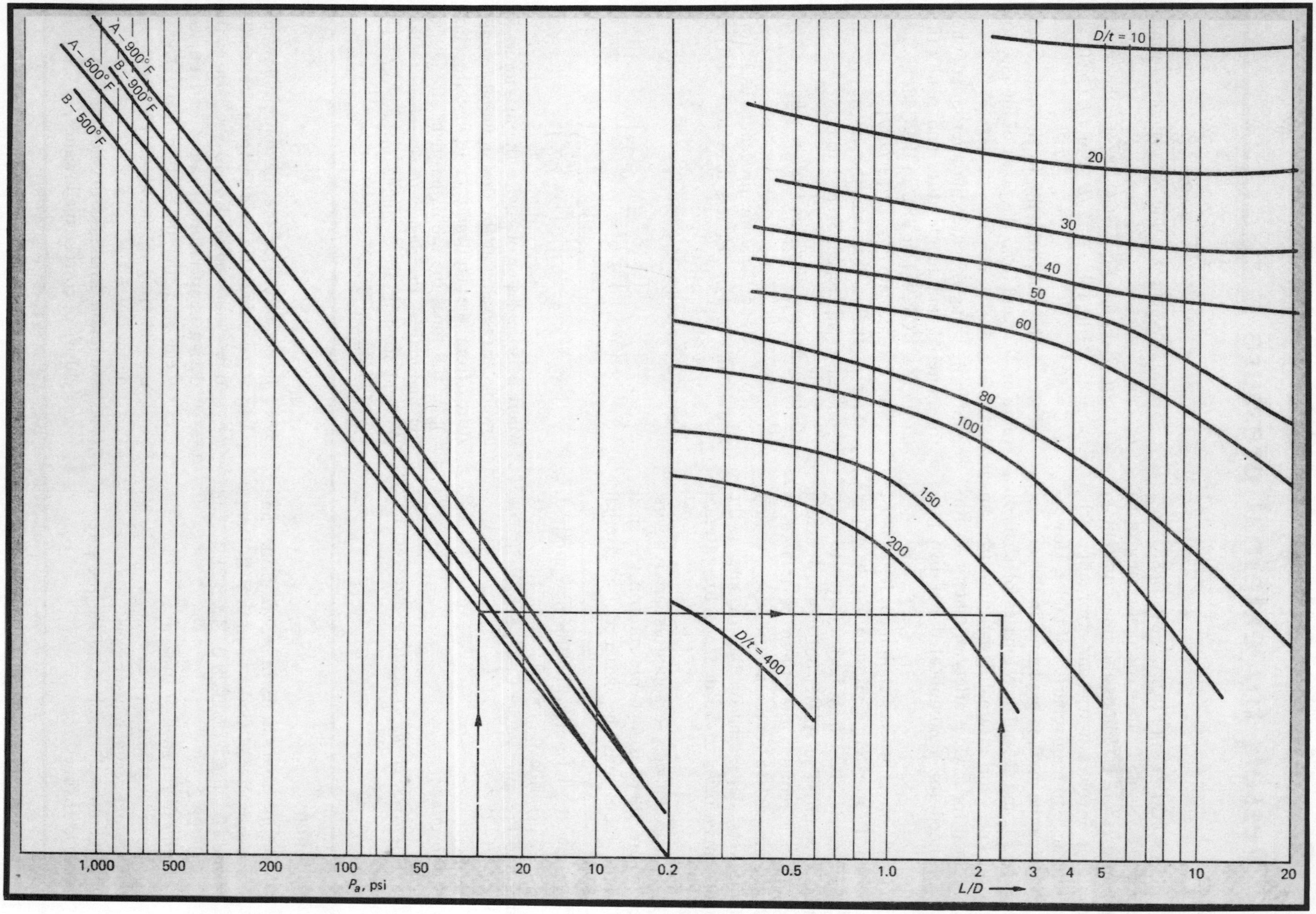


Fig. 1. Chart used for solving problems concerning shells and tubes subjected to external pressure. Use this chart when $D/t > 10$.

Step 3—Choose the smaller $P_{a1}$ and $P_{a2}$ for maximum allowable design pressure. This can be compared with $P$ and $t$, and may be altered if necessary.

### *A Quick Estimating Method*

Figs. 1 and 2 have been developed based on detailed calculations, and give a very good estimate of $P_a$ if $D$, $t$ and $L$ are known. Or, $t$ may be found if the other parameters are known. The two examples that follow are worked out using both methods, to illustrate the versatility of the charts.

Two categories of materials commonly used are considered in these charts under Type A and Type B. They are categorized in Table 1.

### *Example 1*

A shell of material SA 515 Grade 70 is to be designed to withstand external pressure under the following conditions:

$P=30$ psi; $D=80$ in.; $L=180$ in.; design temperature, $T_{,}=500°$F. Estimate the thickness that is suitable.

*Method A (using Code procedure):*

Assume the thickness to be 1.5 in. Then:

$$L/D = 2.25; D/t = 53$$

From Fig. UGO-28, for $L/D=2.25$ and $D/t=53$, find $A=0.0015$.

From Fig. UCS-28.2, for $A=0.0015$ and $T=500°$F, find $B=11{,}000$. Then:

$$P_a = 4B/[3D/t)] = (4)(11{,}000)/(3)(53) = 277 \text{ psi}$$

Therefore, we have to try a lower value for $t$. Try $t=0.5$ in.

Then, $L/D = 2.25; D/t = 160$.

From Fig. UGO-28, $A = 0.00028$.

From Fig. UCS-28.2, $B = 3{,}800$.

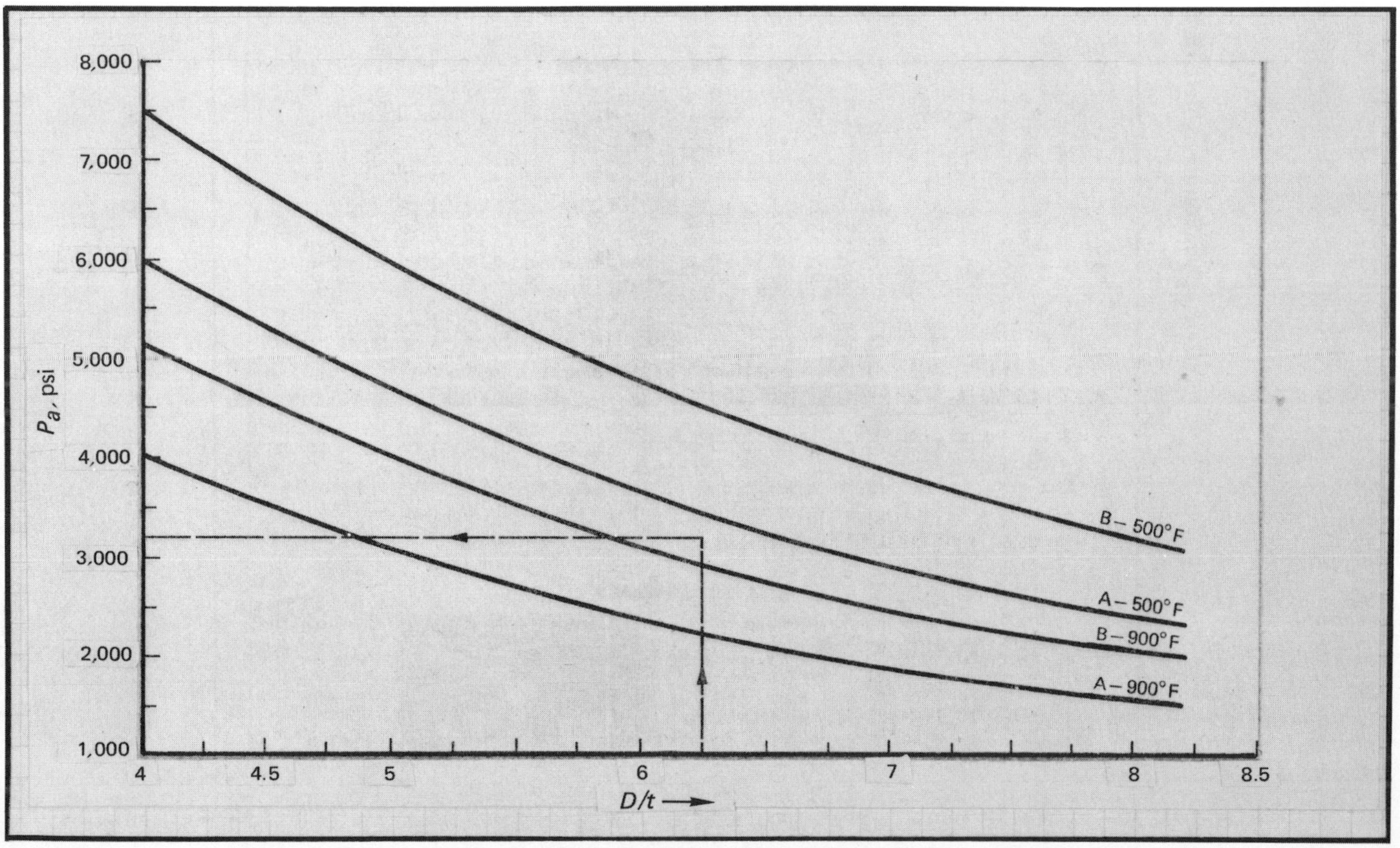


**Fig. 2. Chart for shells and tubes subjected to external pressure. Use when $D/t<10$.**

**Table 1. Classification of Materials for Use in Figs. 1 and 2**

| No. | Material | Grade | Type for Use in Figs. 1 and 2 | No. | Material | Grade | Type for Use in Figs. 1 and 2 | No. | Material | Grade | Type for Use in Figs. 1 and 2 |
|---|---|---|---|---|---|---|---|---|---|---|---|
| 1. | SA 36 | – | B | 24. | SA 178 | A and C | B | 47. | SA 213 | All | B |
| 2. | SA 283 | A and B | A | 25. | SA 226 | – | A | 48. | SA 333 | 4, 7 and 3 | B |
| 3. | SA 283 | C and D | B | 26. | SA 334 | 1 | A | 49. | SA 334 | 7 and 3 | B |
| 4. | SA 285 | A and B | A | 27. | SA 334 | 6 | B | 50. | SA 335 | All | B |
| 5. | SA 285 | C | B | 28. | SA 557 | A | A | 51. | SA 369 | All | B |
| 6. | SA 414 | A and B | A | 29. | SA 557 | B | B | 52. | SA 423 | 1 and 2 | B |
| 7. | SA 414 | C, D, E and F | B | 30. | SA 587 | – | B | 53. | SA 250 | T1 b | A |
| 8. | SA 442 | 55 and 60 | B | 31. | SA 105 | – | B | 54. | SA 250 | T1, T1 a | B |
| 9. | SA 455 | All | B | 32. | SA 181 | 1 and II | B | 55. | SA 334 | 7 and 3 | B |
| 10. | SA 515 | All | B | 33. | SA 216 | WCA and WCB | B | 56. | SA 423 | 1 and 2 | B |
| 11. | SA 516 | All | B | 34. | SA 266 | All | B | 57. | SA 217 | WCI | B |
| 12. | SA 537 | Up to 2.5 in. | | 35. | SA 350 | LF1 and LF2 | B | 58. | SA 336 | Cl F22 a | B |
| | | incl Cl. 1 | B | 36. | SA 352 | LCB | B | 59. | SA 336 | Cl F21 a | B |
| 13. | SA 53 | A and B | B | 37. | SA 372 | I | B | 60. | SA 336 | Cl F5 | B |
| 14. | SA 106 | A, B and C | B | 38. | SA 203 | A and D | B | 61. | SA 350 | LF4, LF3 | B |
| 15. | SA 179 | – | A | 39. | SA 204 | A | B | 62. | SA 352 | LC1 | B |
| 16. | SA 192 | – | A | 40. | SA 387 | 2 Cl. 1 | B | 63. | SA 508 | 1 | B |
| 17. | SA 333 | 1, 6 | B | 41. | SA 387 | 12 Cl. 1 | B | 64. | SA 541 | 1 | B |
| 18. | SA 334 | 1, 6 | B | 42. | SA 387 | 11 Cl. 1 | B | 65. | SA 240 | 405, 410S, 410 | B |
| 19. | SA 524 | I and II | B | 43. | SA 387 | 22 Cl. 1 | B | 66. | SA 268 | TP 405, TP 410 | B |
| 20. | SA 556 | A2 | A | 44. | SA 387 | 21 Cl. 1 | B | 67. | SA 479 | 405 | B |
| 21. | SA 556 | B2 | B | 45. | SA 199 | All | A | | | | |
| 22. | SA 53 | A and B | B | 46. | SA 209 | All | B | | | | |
| 23. | SA 135 | A and B | B | | | | | | | | |

$$P_a = (4)(3{,}800)/(3)(160) = 32 \text{ psi}$$

Hence, $t = 0.5$ is reasonable.

*Method B (using Figs. 1 and 2):*

Since $D/t > 10$, use Fig. 1. From Table 1, SA 515 Grade 70 comes under Type B.

On Fig. 1, start from $P_a = 30$, and go up to cut $T = 500°F$. Move horizontally to intersect a vertical line from $L/D = 2.25$ at a $D/t$ of approximately 160. Hence, $t = D/160 = 0.5$ in.

### *Example 2*

An SA 335 P11 pipe having a dia. of 5 in., a thickness of 0.8 in. and a length of 40 in. is used at a temperature of 800°F. Determine the maximum external pressure that this pipe can withstand.

*Method A (using Code procedure):*

$$L/D = 40/5 = 8; D/t = 5/0.8 = 6.25$$

From Fig. UGO-28, $A = 0.3$.
From Fig. UCS-28.2, $B = 12{,}000$.

$$P_{a1} = [(2.167/6.25) - 0.0833]\,12{,}000 = 3{,}160 \text{ psi}$$

To find $P_{a2}$, we must first obtain $S$.

For the material specified, the allowable stress (from Table UCS-23, in Section VIII, Div. 1) is 15,000 psi, and yield strength (Section VIII, Div. 2) is 22,500 psi.

$$0.9(22{,}500) = 20{,}250$$

Twice the allowable stress is: 2(15,000)= 30,000.

We choose the lower of the two (20,250) to use in the equation for $P_{a2}$.

$$P_{a2} = [(2)(20{,}500)/6.25]\,[1 - (1/6.25)] = 5{,}443 \text{ psi}$$

Hence the maximum allowable pressure is 3,160 psi.

*Method B (using Figs. 1 and 2):*

Since $D/t < 10$, use Fig. 2. From Table 1, material used falls under Type B.

On Fig. 2, for $D/t = 6.25$ and $T = 800°F$, read off $P_a = 3{,}200$ psi.

### *Conclusion*

These charts are reasonably accurate for engineering purposes. We have worked out over 20 cases, and the maximum error was 6%.

For this approach:

- No trial and error is involved if the thickness is not known. No "feel" for the value of $t$ is needed.
- There is no need to spend time looking up values for yield strength and allowable stress at the appropriate temperature.
- There is no need to read values for $A$ and $B$ from the figures in the Code manual (Fig. UGO-28.0 and UGO-28.2); this saves further time.

In addition, it should be pointed out that the two figures in the Code (UGO-28.0 and 28.2) have log-log scales, and the accuracy obtained by their use depends in part on the skill of the user.

From the foregoing examples, one can see the elegance, simplicity and versatility that make the use of Figs. 1 and 2 preferable to the Code methods.

# 8. Maximum allowable working pressures for steel piping

HERE IS A METHOD TO ESTIMATE THE MAXIMUM allowable pressure for ferrous piping according to ASME and ANSI codes without referring to the actual codes' stress tables. In addition, specific pipe dimensions (O.D. and thickness) need not be known, as the schedule number and pipe material are all that are necessary.

Table 1 gives the maximum allowable working pressure, in psig, based on an allowable stress of 15,000 psi. To correct for the specific material and temperature, multiply the value obtained from Table 1 by the appropriate $F$ factor from the figure. Table 2 lists the various materials by composition and ASME code specification.

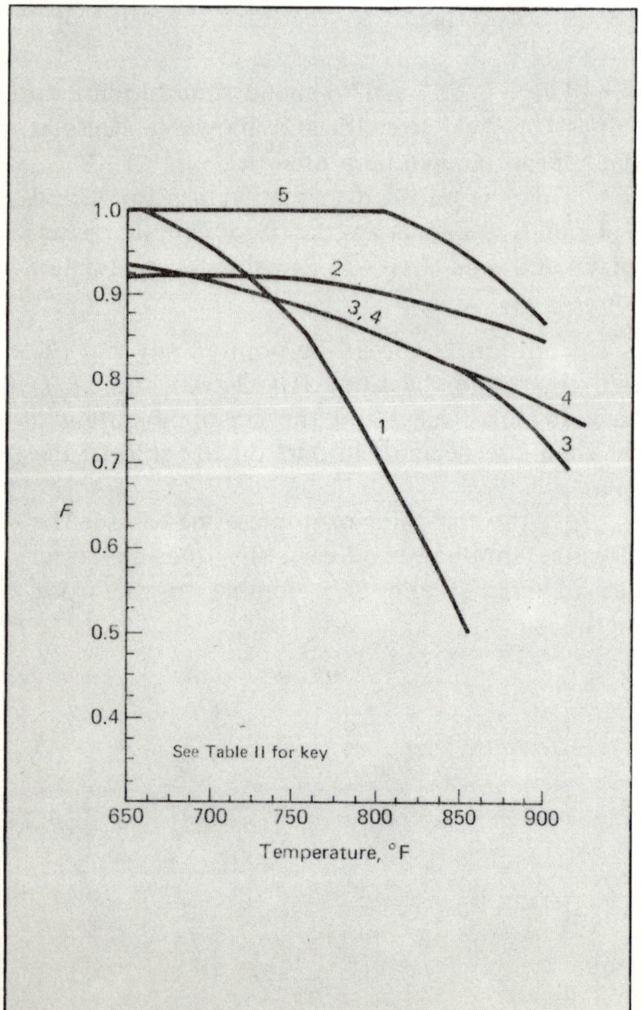


### *Example*

Estimate the maximum allowable working pressure for a 2-in., Schedule 40 carbon-manganese (SA 53B) pipe at 750°F.

### *Solution*

From Table 1, a value of 1,782 psi is obtained. From the figure, for SA 53B (Curve 1) at 750°F, $F$ is 0.86. Hence, the maximum allowable pressure is $0.86 \times 1{,}782 = 1{,}532$ psig.

**Table 1. Maximum Allowable Pressure***

| *Nominal Pipe Size, in.* | *Schedule 40* | *Schedule 80* | *Schedule 160* |
|---|---|---|---|
| ¼ | 4,830 | 6,833 | – |
| ½ | 3,750 | 5,235 | 6,928 |
| 1 | 2,857 | 3,947 | 5,769 |
| 1½ | 2,112 | 3,000 | 4,329 |
| 2 | 1,782 | 2,575 | 4,225 |
| 2½ | 1,948 | 2,702 | 3,749 |
| 3 | 1,693 | 2,394 | 3,601 |
| 4 | 1,435 | 2,074 | 3,370 |
| 5 | 1,258 | 1,857 | 3,191 |
| 6 | 1,145 | 1,796 | 3,076 |
| 8 | 1,006 | 1,587 | 2,970 |

*Based on allowable stress of 15,000 psi

**Table 2. ASME Specifications for Various Steels**

| *Composition* | *Code Designation* | *Curve on Figure* |
|---|---|---|
| Carbon-manganese | SA 53B | 1 |
| Carbon-manganese-silicon | SA 106B | 1 |
| Carbon-½Mo (½% molybdenum) | SA 335 P1 | 2 |
| 5Cr-½Mo (5% chromium, ½% molybdenum) | SA 335 P5 | 3 |
| 9Cr-1Mo | SA 335 P9 | 4 |
| 1¼Cr-½Mo | SA 335 P11 | 5 |
| 2¼Cr-1Mo | SA 335 P22 | 5 |

Source: ASME boiler and pressure vessel code, Sec. 1, 1980, p. 184, Table PG23.1

# 9. Chart aids estimation of pressure vessel weight

THE WEIGHT OF PRESSURE VESSELS IS AN IMportant factor for engineers engaged in design and production. It gives an idea of the required material-handling facilities and of cost. Consequently a convenient method for bypassing slide-rule or desk-computer calculations for estimating vessel weights might be useful. This nomograph provides one such method, and allows designers to make visual comparisons.

The "Chemical Engineers' Handbook," section 24, gives the following formula for the weight of a pressure vessel:

$$W = 136\, D(H + 0.8D)(t + X)$$

where:

$W$ = weight, lb
$D$ = diameter, ft
$H$ = length of shell, ft
$t$ = average shell thickness, in.
$X$ = an equivalent thickness, in., to be allowed for nozzles, manholes, etc., and equals 0.08 for vessels with no internals and 0.15 for vessels with occasional manholes.

The nomograph gives $W$, if the other variables are known.

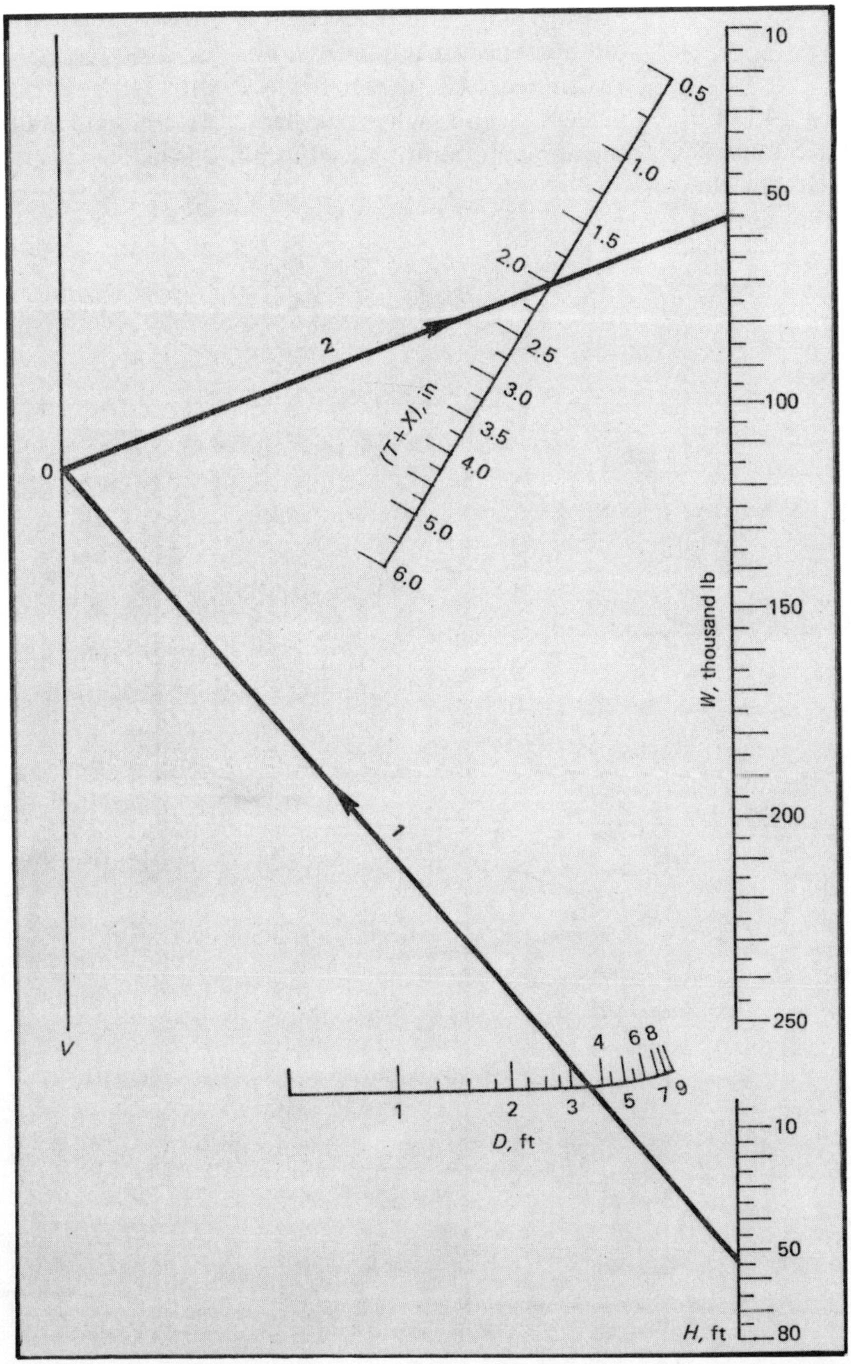


### *Example*

If $D = 3.5$ ft, $H = 55$ ft, $t = 1.92$ in. and $X = 0.08$, what is $W$?

### *Solution*

On the nomograph, connect 55 on $H$ with 3.5 on $D$ and extend a straight line to reference line $V$; join this point with $(t + X) = 2.00$ and read 55,000 lb on the $W$ scale.

# 10. Chart determines fan horsepower and annual operating cost

THIS NOMOGRAPH PROVIDES A SIMPLE METHOD for determining horsepower requirements and annual operating costs for industrial fans. Air flow rate, static pressure, fan efficiency, and electrical cost must be known to use the nomograph.

### *Example*

A 70 percent efficient fan is handling 145,000 cfm of air against a static pressure of 1.5 in. w.g. using electricity costing 4¢/kwh. What are the fan horsepower requirement and the annual operating cost?

### *Solution*

Align 1.5 in. w.g. on Scale S with 145,000 cfm on Scale Q and extend the line to the reference line. Connect the marked point on the reference line with 70 percent efficiency on Scale E and, where Scale P is crossed, read the fan horsepower requirement as 48.

Connect 48 hp on Scale P with 4¢/kwh on Scale $C_e$, and, where Scale $C_o$ is crossed, read the annual operating cost as $130,000.

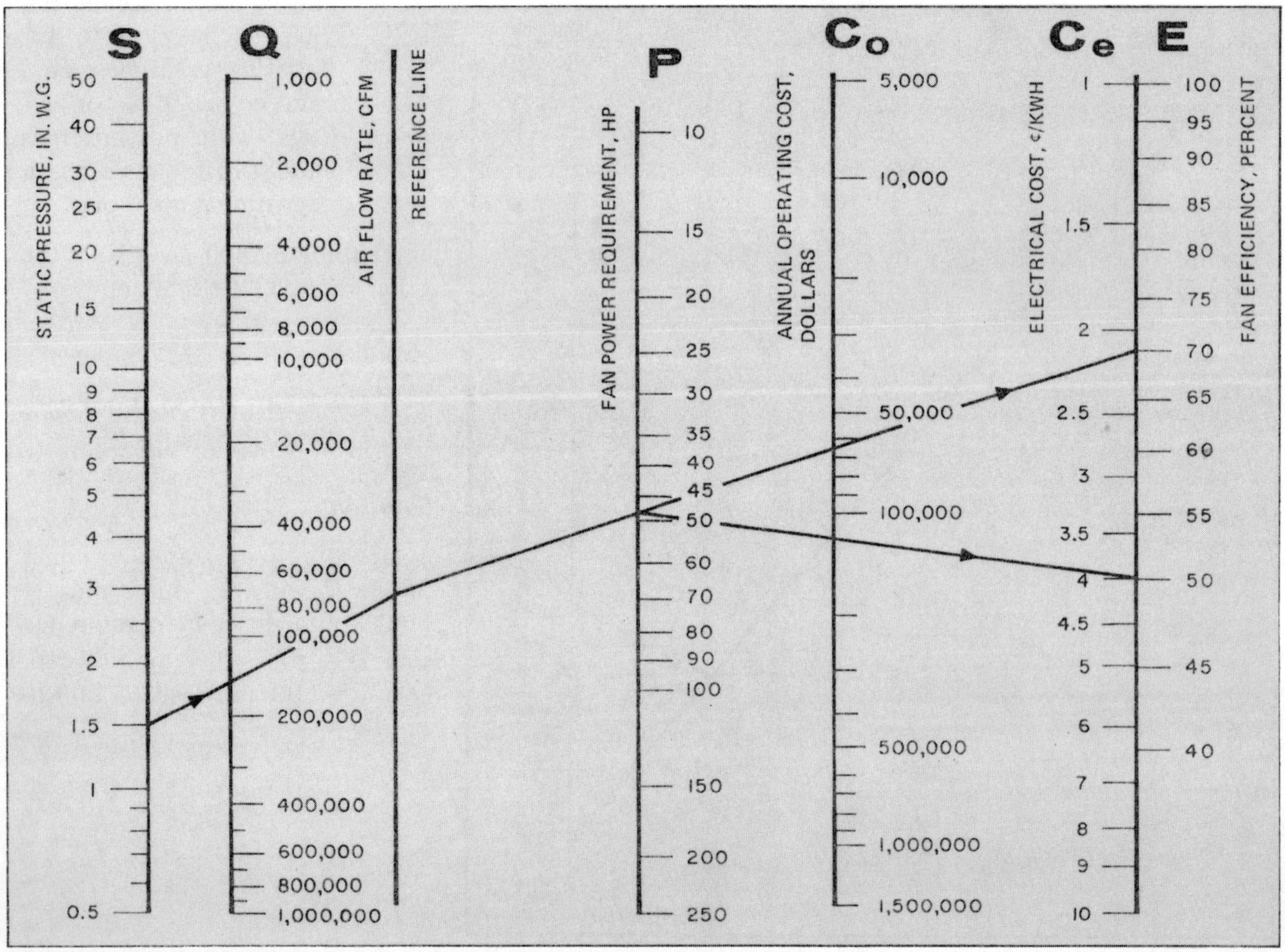

# 11. Chart provides quick estimate of pumping variables

IN THE ANALYSIS OF ANY PUMPING SYSTEM there are four variables that must be considered: head, volumetric flow, pressure, and density of the pumped fluid. It is often helpful when a quick conversion between two values or the relationship of one variable to another can be quickly determined. This nomograph can be used to solve the equations:

$$P=\frac{0.5\times 10^{-6}\,WH}{E}$$

where:

$P$ = power required to drive pump, hp
$W$ = mass flow of liquid, lb/hr
$H$ = head, ft of liquid
$E$ = overall efficiency of pump with drive, decimal

$$Q=\frac{W}{8\rho}$$

where:

$Q$ = volumetric flow, gpm
$\rho$ = density, lb/cu ft

$$p=\frac{\rho H}{144}$$

where:

$p$ = pressure, psi

***Example***

A pump delivers 200 gpm of a liquid having a density of 63 lb/cu ft and a head of 350 ft. Assume efficiency of the pump with driver is 75 percent; determine the mass flow, pressure, and power required to drive the pump.

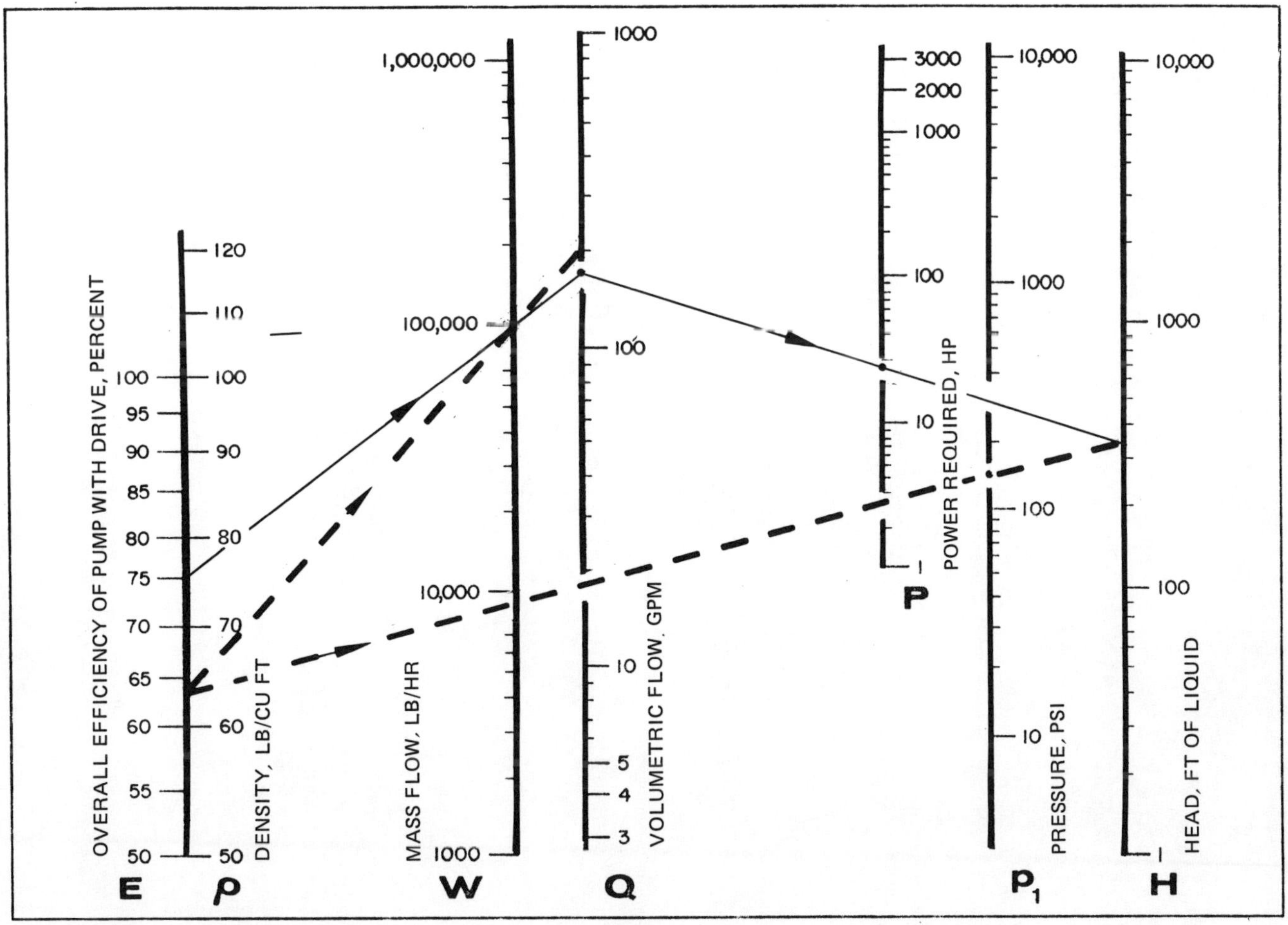

***Solution***

Connect 63 lb/cu ft on Scale $\rho$ with 200 gpm on Scale Q, and, where Scale W is crossed, read mass flow as 100,000 lb/hr. Connect 75 percent on Scale E with 100,000 lb/hr on Scale W, and extend the line to intersect Scale Q; then, connect this point on Scale Q with 350 ft on Scale H, and where Scale P is crossed, read power requited as 25 hp. Then, connect 63 lb/cu ft on Scale $\rho$ with 350 ft on Scale H, and where Scale p is crossed, read pressure as 150 psi.

The chart can be used in reverse to determine head $H$ if pressure $p$ and density $\rho$ are known. Similarly, if mass flow $W$, pressure $p$, and head $H$ are given, other parameters–volumetric flow $Q$, power $P$, and density $\rho$–can be found.

# 12. Charts for evaluating pump performance from motor readings

PLANT ENGINEERS FREQUENTLY NEED TO CHECK the performance of a pump to determine its efficiency for comparison with the predicted performance, or to establish the rate of flow or head being developed.

Pump data related to motor readings of current and voltage may be more accurate as an indication of pump flow than direct measurement using an orifice meter that is choked or improperly calibrated. In general, electric motor efficiency and power factor are more constant than pump efficiency over the normal range of pump operation.

Pump power consumption is given by

$$P = \frac{0.746\, QHS}{3960\, E_p\, E_m} \quad (1)$$

where:

$P$ = power, kw
$Q$ = pump flow, gpm
$H$ = pump head, ft of liquid
$\Delta p$ = pump differential pressure, psi
$S$ = specific gravity of liquid pumped
$E_p$ = pump efficiency, fraction
$E_m$ = motor efficiency, fraction

Pump head is given by

$$H = \frac{2.3\, \Delta p}{S} \quad (2)$$

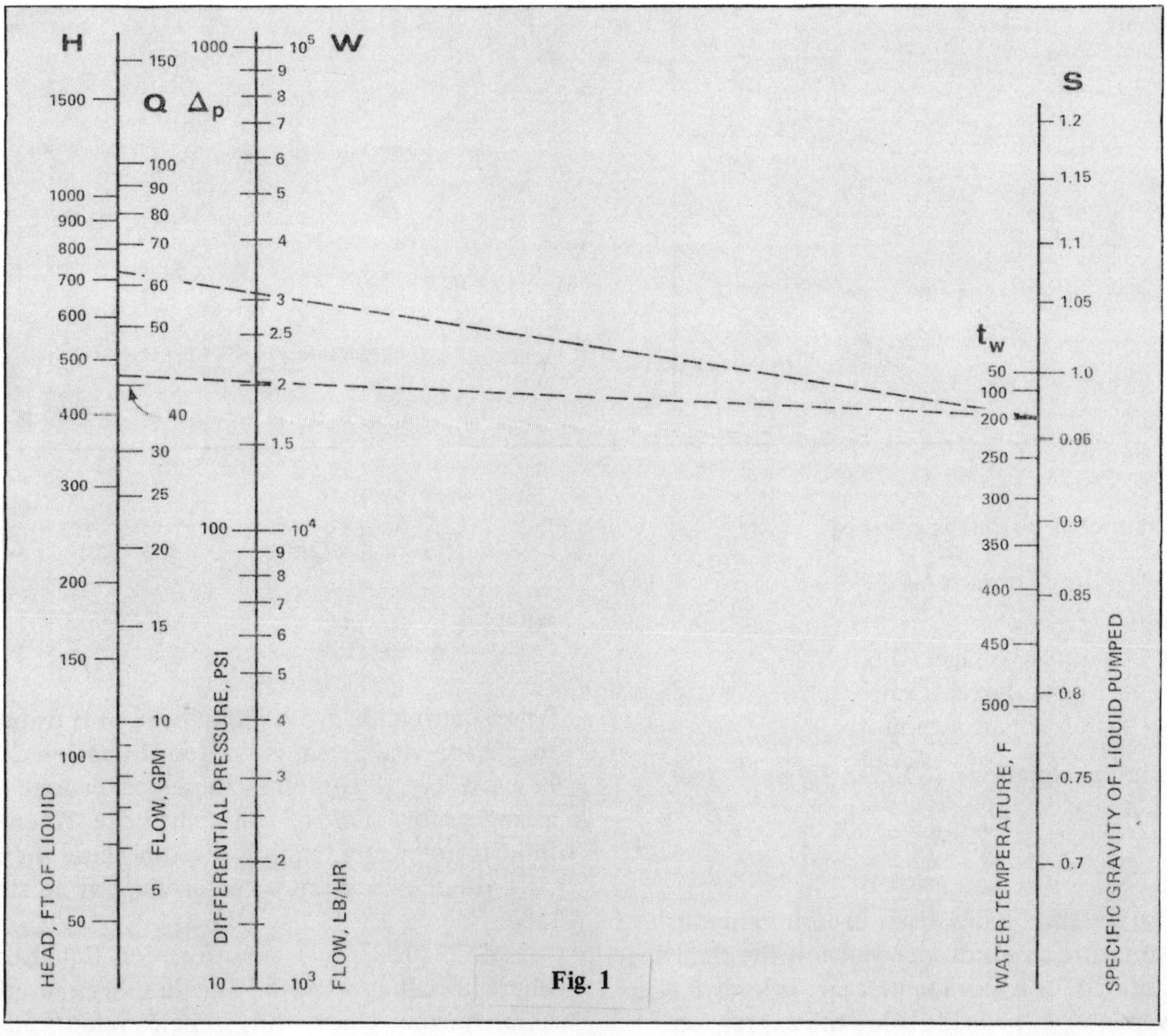


Fig. 1

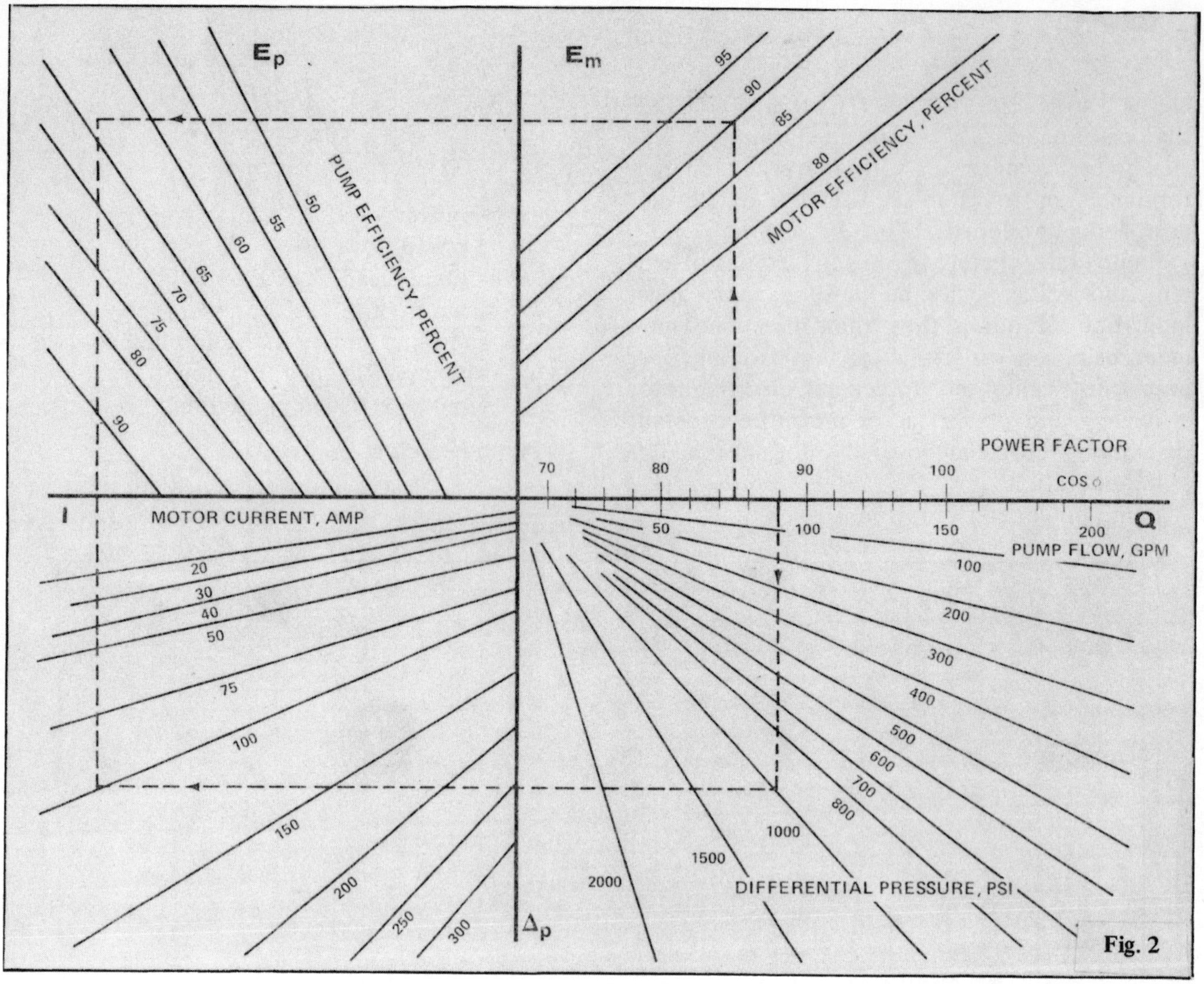


Fig. 2

The motor output is given by

$$P = \sqrt{3}\, VI \cos \phi \qquad (3)$$

where:

$V$ = motor voltage, volts
$I$ = motor current, amp
$\cos \phi$ = power factor of motor

Combining equations 1, 2, and 3 yields

$$Q = \frac{1800\, E_p\, E_m\, \cos \phi I}{\Delta p} \qquad (4)$$

Flow is sometimes given in lb/hr and must be converted to gpm, or vice versa, if the fluid specific gravity or water temperature is known. This conversion can be made using the expression:

$$Q = \frac{W}{500S} \qquad (5)$$

where:

$W$ = pump flow, lb/hr

When conversion from pump head in ft of liquid to psi, or vice versa, is needed, Equation 2 can be used. Fig. 1 combines these conversions with a single movement of a straight edge. When the fluid is hot water, time is saved because the specific volume need not be determined from steam tables.

Fig. 2 provides a solution for Equation 4 where a value of 460 volts has been used to simplify the chart. Any single value can be

**TYPICAL MOTOR EFFICIENCY, POWER FACTOR DATA**

| *Motor Size HP* | *Motor Efficiency, Percent* | | *Power Factor, Percent* | |
|---|---|---|---|---|
| | *Full Load* | *Half Load* | *Full Load* | *Half Load* |
| 5 | 86.5 | 83.7 | 86.6 | 72.5 |
| 10 | 88.5 | 85.2 | 83.6 | 68.0 |
| 20 | 88.5 | 87.2 | 86.9 | 79.7 |
| 40 | 90.2 | 87.9 | 85 | 73.8 |
| 50 | 91.7 | 89.5 | 86.1 | 76.3 |
| 75 | 91.7 | 88.4 | 82.8 | 68.0 |
| 100 | 93.0 | 91.0 | 89.9 | 82.8 |
| 200 | 94.1 | 92.3 | 91.5 | 86.1 |
| 300 | 94.1 | 92.7 | 90.2 | 84.7 |

Source: Extracted from Westinghouse Electric Corp. motor catalog (1800 rpm).

determined when the remaining values are given. If flow $Q$ is known, the pump efficiency can be determined if motor efficiency $E_m$, power factor for motor cos $\phi$, motor current $I$, and pump differential pressure $\Delta p$ are known. The table provides a good estimate of typical $E_m$ and cos $\phi$ values.

Using pump characteristic curves, the plant engineer can infer whether the pump performance is satisfactory or substandard. Where the result is far away from the predicted efficiency, a double check of the flow meter for choking, plugging, or calibration is indicated. If the efficiency varies by a wide margin, the pump bearings or impeller may be worn or misaligned. This condition is often confirmed by vibration.

Whatever the cause, these charts can be used to help the plant engineer keep tabs on pump performance. Even when there is close agreement between predicted and measured data, this method serves as a double check on pump performance.

### Example

In a boiler feedwater system 30,000 lb/hr of water at 200F is pumped at 200 psi. Determine the flow in gpm and the head in ft of water.

### Solution

On Fig. 1, connect 200F on Scale $t_w$ with 30,000 lb/hr on Scale $W$ and extend to intersect Scale $Q$ at 62 gpm. Connect 200F on Scale $t_w$ with 200 psi on Scale $\Delta p$ and extend to intersect Scale $H$ at 475 ft. Scales $\Delta p$, $H$, $Q$, and $W$ have the same multiplication factor; hence the range of the chart is very wide.

### Example

A 460v 100 hp electric motor drives a pump that delivers 90 gpm with a differential pressure $\Delta p$ of 1000 psi. The current draw is 100 amp. Determine efficiency of the pump.

### Solution

From the table select motor efficiency of 90 percent and power factor of 85 percent. In Fig. 2, starting at 85 percent on Scale cos $\phi$, extend a line up to intersect with 90 percent on Scale $E_m$; then extend a line through Scale $E_p$. From 90 gpm on Scale $Q$ extend a line to intersect Scale $\Delta p$ at 1000 psi. From the point of intersection, extend a line left to intersect with 100 amp on Scale $I$. From that intersection extend a line into Scale $E_p$ to intersect the horizontal line and read pump efficiency as 65 percent.

# 13. How to evaluate heat-exchanger bids using life cycle costing methods

A HEAT EXCHANGER CAN BE DESIGNED FOR the same duty conditions in a variety of ways. By using large flow velocities for both the media, the size or the capital cost can be brought down significantly. However, the operating costs increase as the pressure drop of the fluids increases.

**Table 1. Data on Exchanger Bids**

| *Item* | *Bid 1* | | *Bid 2* |
|---|---|---|---|
| Shell side fluid | | Kerosine | |
| Tube side fluid | | Water | |
| Shell side flow, lb/hr | | 45,000 | |
| Tube side flow, lb/hr | | 15,000 | |
| Duty, MBtu/hour | | 5.3 | |
| Surface area, ft$^2$ | 700 | | 630 |
| Shell side pressure drop, psi | 9 | | 11.5 |
| Tube side pressure drop, psi | 4 | | 7.5 |
| Density of kerosine, lb/cu ft | | 45 | |
| Density of water, lb/cu ft | | 56 | |
| Cost of exchanger with controls, supports, insulation, $ | 14,700 | | 13,000 |

*Nomenclature*

$C_a$ = Annual cost of operation, $
$C_e$ = Cost of electricity, $/kwh
$C_o$ = Capitalized cost of operation, $
$C_T$ = Total capitalized cost, $
$e$ = Escalation rate
$F$ = Capitalization factor
$i$ = Interest rate
$I$ = Investment on equipment, $
$P$ = Additional power consumption, kw
$T$ = Life of the equipment, years
$W$ = Flow through shell and tube side, lb/hr (subscript $S$ refers to shell and subscript $t$ to tube)
$\rho$ = Density of fluid, lb/cu ft
$\eta$ = Efficiency of pump-drive system

This is a recurring expenditure and has to be reckoned with throughout the life of the equipment. By investing a lesser amount on the smaller heat exchanger, you may add to the operating costs.

How does one evaluate, then, whether the equipment selected is optimum for the duty conditions, given a number of alternatives?

Life cycle costing techniques may be applied in such cases. The sum of the capitalized cost of operation and the capital cost of the equipment indicates the economics of the various alternatives.

In what follows, an example illustrates how one arrives at the total capitalized cost, which is an index of the competitiveness of the bid.

### *Operating Costs*

For a heat exchanger, the major operating cost is the pumping power. For example, in a shell and tube exchanger, fluids are pumped through the tubes and over them. The pumping power $P$ is given by the following formula for liquids:

$$P = 5.4 \times 10^{-5} \frac{W \Delta P}{\rho \eta} \qquad (1)$$

By knowing the period of operation and cost of electricity, it is possible to compute the annual additional cost of pumping the fluid through the heat exchanger, $C_a$:

$$C_a = P\, C_e\, N \qquad (2)$$

The next step is to compute the capitalized cost of operation.

$$C_o = C_a F \qquad (3)$$

where $F$, the capitalization factor is given by:

$$F = \left(\frac{1+e}{1+i}\right) \left[ \frac{1 - \left(\frac{1+e}{1+i}\right)^T}{1 - \left(\frac{1+e}{1+i}\right)} \right] \qquad (4)$$

In this equation, $e$ is the escalation in the cost of electricity, and $i$, the interest rate.

The total capitalized cost is:

$$C_T = (C_o + I) \tag{5}$$

where $I$ is the equipment cost or investment.

The alternative with the last $C_T$ is the best one, assuming all are technically acceptable.

***Example***

Two bids were received for a shell and tube heat exchanger for the same duty. The parameters are given in Table 1. The customer has to choose between them.

***Solution***

Various parameters have to be obtained before a preliminary evaluation can be made. They are:

$i$, interest rate = 14%
$e$, escalation rate = 8%
$N$, period of operation = 6,000 hrs/yr
$T$, life of the equipment = 15 years
$C_e$, cost of electricity = 40 mills/kwh

First compute $F$:

$$F = \left(\frac{1.08}{1.13}\right)\left[\frac{1-\left(\frac{1.08}{1.13}\right)^{15}}{1-\left(\frac{1.08}{1.13}\right)}\right] = 10.64$$

Pumping power $P$ has to be computed. For Bid 1, shell side power requirement, assuming 75% pump drive efficiency, is:

$$P_s = \frac{5.45 \times 10^{-5} \times 45{,}000 \times 9}{45 \times 0.75}$$
$$= 0.654 \text{ kw}$$

$$P_t = \frac{5.45 \times 10^{-5} \times 15{,}000 \times 4}{56 \times 0.75}$$
$$= 0.78 \text{ kw}$$

Annual operating cost, $C_a = (0.654 + 0.78) \times .04 \times 6{,}000 = \$344$.

Capitalized cost of operation $= C_o = 344 \times 10.64 = \$3{,}662$.

$C_T$, total capitalized cost of Bid 1 = (14,700 + 3,662) = \$18,362. Similarly, for Bid 2, $C_T$ = \$18,869.

Thus, it appears that Bid 2, though lower in capital cost of equipment, has a higher total capitalized cost. Hence, Bid 1 is cheaper and may be chosen if technical requirements, such as space, layout, design, etc., are acceptable.

A different picture emerges when the cost of electricity or period of operation changes. Let the period of operation be only 3,000 hours. Then, for Bid 1:

$$C_a = 344 \times \frac{3{,}000}{6{,}000} = \$172$$

$$C_o = 172 \times 10.64 = \$1{,}830$$

$$C_T = (1{,}830 + 14{,}700) = \$16{,}530.$$

Similarly for Bid 2, $C_T$ = \$15,932.

Bid 2 becomes a better choice now.

In this way, engineers may evaluate bids considering operating costs.

Such analysis is warranted these days due to the increased cost of energy.

# 14. Selecting optimum steam line size using life cycle costing methods

*Nomenclature*

$e$ = escalation rate
$E$ = efficiency of joint
$f$ = friction factor
$F$ = capitalization factor
$i$ = interest rate
$d$ = outer diameter of pipe, in.
$d_1$ = inner diameter of pipe, in.
$N$ = additional power required by boiler feed pump, kw
$P$ = design pressure of pipe, psig
$C_a$ = annual cost of operation, $
$C_e$ = total capitalized cost, $
$S$ = allowable stress per code, psi
$L_e$ = total equivalent of pipe, ft
$t$ = thickness of pipe, in.
$v$ = specific volume of steam, cu ft/lb
$\Delta P$ = steam line pressure drop, psi
$\rho w$ = density of water at feed pump, lb/cu ft
$V$ = line velocity, fps
$\eta$ = efficiency of pump drive system
$Ny$ = life of the plant, years

PIPING ACCOUNTS FOR AS MUCH AS 30% OF THE total plant costs and selection of sizes must be made with care. Though general recommendations are available for line velocities for various fluids, engineers would do well to make some preliminary analysis before arriving at a particular size.

Too high a line velocity results in a high pressure drop, which means large operating costs. Computed over the life of the plant, this could be a substantial expenditure.

On the other hand, if a low line velocity is used a large pipe has to be chosen. This means higher material cost, higher insulation cost, larger fabrication cost, higher erection costs, and higher pipe hanger and support costs.

An optimum line velocity may be determined for each case. The following example illustrates the basics behind choice of an economical steam piping system for a process boiler plant.

### *Example*

Assume 6,000,000 lb/hr of steam is required at 800 psia and 900°F in a process plant. Design pressure at the exit of the boiler superheater is 1,000 psia, thus a maximum drop of 200 psi may be tolerated in the piping between the boiler and process steam header.

**Table 1. Study of Optimum Pipe Size**

| *d, in.* | *t,* in.* | *f* | *V, fps* | $L_e$*, ft* | $\Delta P$*, psi* | *N, kw* | $C_a \times F$ *$* | *Fixed Cost, $†* | $C_e$*, $* |
|---|---|---|---|---|---|---|---|---|---|
| 8.625 | 0.406 | 0.0141 | 465 | 212 | 115 | 100 | 317,100 | 17,000 | 334,100 |
| 10.75 | 0.50 | 0.0137 | 299 | 248 | 43.4 | 38 | 120,498 | 23,000 | 143,498 |
| 12.75 | 0.562 | 0.0133 | 210 | 272 | 19.1 | 17 | 53,273 | 29,000 | 82,273 |
| 14 | 0.593 | 0.012 | 173 | 288 | 12.2 | 11 | 33,930 | 35,000 | 68,930 |
| 16 | 0.656 | 0.0127 | 131 | 312 | 6.53 | 6 | 18,138 | 54,000 | 72,130 |
| 18 | 0.75 | 0.0124 | 104 | 328 | 3.74 | 3.3 | 10,400 | 65,000 | 76,400 |

*Thickness used is the nearest available thickness to that required by ASME code.
†Includes cost of pipe, insulation, valves, pipe hangers, fabrication, and erection.

SA 335 P11 pipes of 8⅝, 10¾, 12¾, 14, 16, and 18 in. OD are available. It is estimated that the developed length of pipe required will be around 100 ft. Also two 90° bends and a swing check valve have to be considered while computing the pressure drop.

Determine the optimum pipe size, if preliminary cost of piping including insulation, hangers, fabrication, and erection is as given in Table 1. The cost of power is 35 mills/kwh. The plant works for 6,000 hr/year and has a life of 20 years. Interest rate is 9% and escalation rate is 6%.

***Solution***

The design pressure of the piping is taken as 1,000 psig and the effect of steam line pressure drop on design pressure is considered marginal.

Also, the impact on boiler costs for small differences in design pressure arising out of line pressure variations will be neglected in this preliminary analysis.

*Step 1.* ASME code formula is used to determine the minimum pipe thickness.

$$t=(P/d)/(2SE+0.8P) \tag{1}$$

$S$, the allowable stress from ASME Boiler and Pressure Vessel Code, Sec. I for SA 335 P11 material at 900°F is 13,100 psi. Using $P=1{,}000$ and $S=13{,}100$, $E=1$. Then $t$ was computed for each pipe and the nearest standard thickness available from the market was chosen.

*Step 2.* Pressure drop for each alternative has to be determined. The following equation was used:[1]

$$\Delta P=3.36\times10^{-6} f w^2 L_e v/d_1{}^5 \tag{2}$$

Friction factor $f$ depends on Reynolds Number which was computed using a chart;[2] $f$ was then looked up from Moody's chart.[3]

The equivalent length has to consider the effect of bends and the swing check valve, the resistances for which were taken from standard tables.[3]

*Step 3.* The pressure drop is then computed. The additional feed pump power associated with the pressure drop for each case is computed using the following equation, assuming 70% efficiency for the feed pump drive.

$$N=\frac{w}{3600}\times\frac{0.746}{550}\times\frac{144}{\eta}\times\frac{\Delta P}{\rho_w}=0.88\Delta P \tag{3}$$

Annual cost of operation is then:

$$C_a=N\times6000\times0.35=210N \tag{4}$$

*Step 4.* The life cycle cost of operation is obtained by multiplying $C_a$ by $F$, where:

$$F=\left(\frac{1+e}{1+i}\right)\left[\frac{1-\left(\frac{1+e}{1+i}\right)^{Ny}}{1-\left(\frac{1+e}{1+i}\right)}\right] \tag{5}$$

For $i=9\%$, $e=6\%$ and $N=20$, $F=15.1$.

Thus, $$C_a\times F=3171N \tag{6}$$

The capitalized cost of operation is then $C_a\times F$.

The total capital cost $C_c$ of the piping system is given by adding the fixed cost with the capitalized operating cost.

$$C_c=(C_a\times F+\text{fixed cost}) \tag{7}$$

The results are summarized in Table 1.

It is seen that the 14-in. OD pipe gives the lowest life cycle cost and hence is recommended. Optimum line velocity is 173 fps.

In practice, a deeper analysis is warranted. For high pressure boilers, pressure drop in steam lines is critical and alters the boiler design pressure.

For pressures in the neighborhood of 2,400 psi, the boiler circulation system is likely to be affected as even a small change affects the process of circulation in the furnace walls. Routing of lines and flexibility analysis should be performed before arriving at the exact length of piping.

Fabrication costs associated with local site conditions and experience would be helpful in arriving at the investment costs on a particular size of pipe.

Nevertheless, this article is aimed at giving an idea of the preliminary analysis involved in arriving at an optimum size of piping. The logic may be applied for large capacity plants, too, where there may be two or even four lines of smaller size instead of one line, as in this case.

### *References*

1. Ganapathy, V., "Chart gives pressure loss in steam lines," OGJ, Jan. 9, 1978, p. 95.
2. Ganapathy, V., "Reynolds Number given for superheated steam," OGJ, Nov. 6, 1978, p. 65.
3. Kern, R., "Practical piping design," *Chemical Engineering,* 1975, reprint.

# 15. Optimum design of waste-heat boilers

WHEN CONFRONTED WITH ALTERNATIVES FOR equipment, engineers should make a selection based on economics, availability, life and ease of operation. More often, engineers choose a design based on nontechnical reasons and justify the choice of parameters as "based on experience."

Here is an analysis of a design of waste-heat boiler where the final choice was based on minimum total capitalization cost. The design procedure is described in detail along with the decision criteria.

### *Example*

A waste-heat boiler is to be designed for the following parameters:

a. Total sulfur gas quantity = 25,000 kg/h
b. Gas composition by volume:
   $SO_2$ = 9 percent
   $O_2$ = 12 percent
   $N_2$ = 79 percent
c. Gas temperature entering the boiler = 1,200°C
d. Gas temperature leaving the boiler = 350°C
e. Steam generation pressure = 25 kg/cm$^2$ (a)
f. Gas properties used[1]
   $C_{pm}$ at 1,200°C = 0.2488 kcal/kg°C
   $C_{pm}$ at 350° = 0.228 "
   $C_{pi}$ at 775°C = 0.263 "
   $\mu$ at 775°C = 0.14 kg/mh
   $k$ at 775°C = 0.053 kcal/mh°C
g. Tubes available: SA/178/gr/A—$\phi$31.8 × 2.9, $\phi$44.5 × 3.2, $\phi$51 × 3.6
h. Shell material, SA 515 gr 70

*Cost and relevant data:*

| | | |
|---|---|---|
| Cost of shell per ton | = | $5,000 |
| Cost of tubes per m$^2$ | = | $ 150 |
| Cost of electricity | = | 30 mills/kwh |
| Escalation rate | = | 8 percent |
| Interest rate | = | 10 percent |

*Assumptions:* General arrangement of the boiler is shown in Fig. 1. Since steam generation is fixed by the gas parameters, the size of the steam drum, down comer and riser system are nearly the same regardless of the size of the fire tube boiler. Hence, optimization is based on the cost of the shell, tubes and power required to move the waste gas through the boiler, which is a very reasonable premise. Hence, sizing of the steam drum and piping are not considered in detail. Cost of maintenance and repairs are considered nearly equal for all the alternatives.

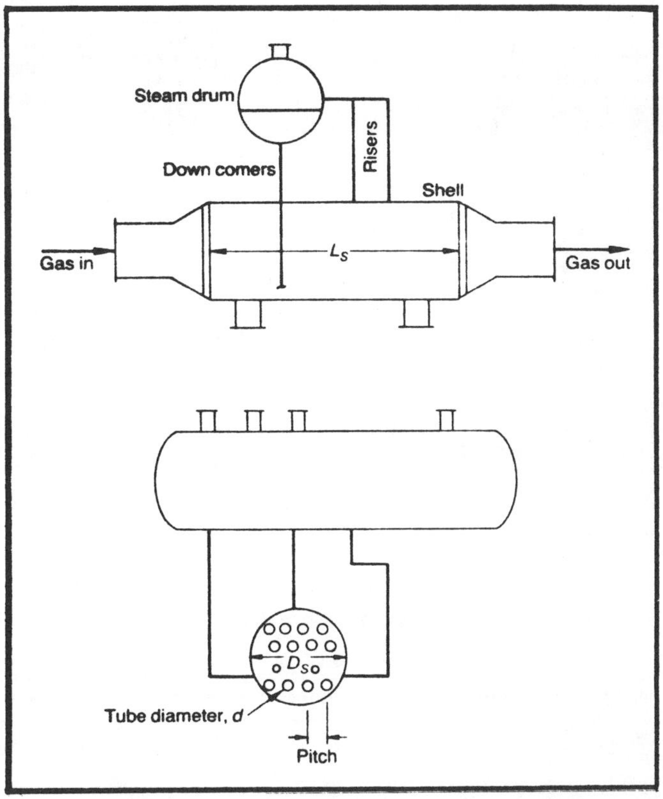


**Fig. 1. Arrangement of waste-heat boiler**

### *Design Method*

Surface area, $A$, of the tubes is derived using the equation:

$$A = Q/U\Delta T = \pi d N L_s \quad (1)$$

(2 percent radiation loss assumed).

$U$, the overall heat transfer coefficient is obtained as follows:

$$U = 0.95 h_{io} \quad (2)$$

Considering the effects of fouling, outside coefficient and metal resistance; $h_i$ is estimated using the correlation:[2]

$$\frac{h_i d_i}{k} = 0.023 \left(\frac{\rho V d_i}{\mu} 3600\right)^{0.8} \left(\frac{\mu C_{Pi}}{k}\right)^{0.4} \quad (3)$$

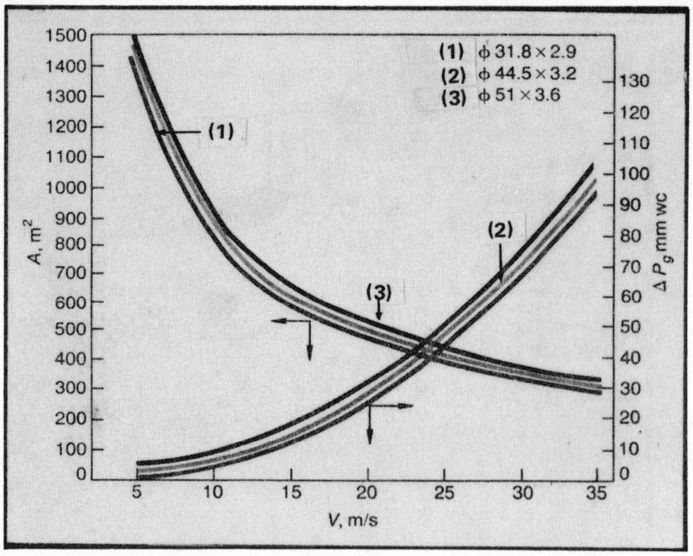


**Fig. 2. Effect of gas velocity on surface area and gas pressure drop**

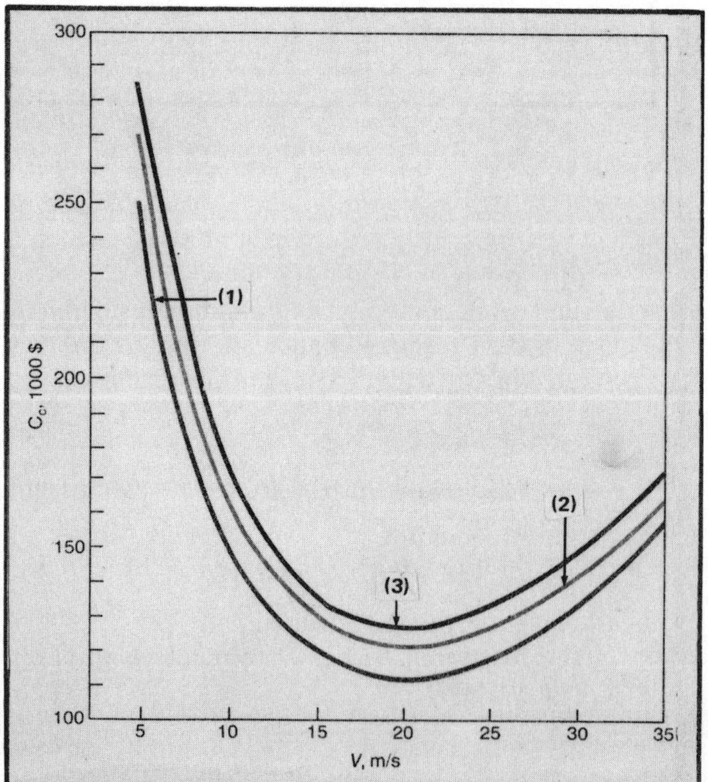


**Fig. 3. Capitalization cost vs. velocity**

### *Nomenclature*

$A$ = Surface area of tubes, $m^2$
$C_o$ = Capitalized cost of operation of boiler, \$
$C_s$ = Cost of shell, \$
$C_t$ = Cost of tubes, \$
$C_{pm}$ = Mean specific heat, kcal/kg°C
$C_{pi}$ = Instantaneous specific heat, kcal/kg°C
$D_s$ = Diameter of shell, m
$d, d_i$ = Tube outer and inner diameters, m
$e$ = Escalation rate, fraction
$F$ = Capitalization cost factor, Equation (15)
$f$ = Friction factor
$g$ = Acceleration due to gravity, $m/s^2$
$h_i$ = Gas side heat-transfer coefficient inside the tubes, $kcal/m^2 h°C$
$h_{io}$ = Corrected coefficient for tube outer surface
$i$ = Interest rate, fraction
$k$ = Gas thermal conductivity, kcal/mh°C
$L_s$ = Length of shell, m
$MW$ = Molecular weight of gas
$N$ = Number of tubes
$N_y$ = Life of equipment, years
$Q$ = Duty of the boiler, kcal/h
$t_g$ = Average gas temperature, °C
$t_1, t_2$ = Gas temperature entering and leaving the boiler, °C
$t_s$ = Shell thickness, m
$\Delta T$ = Log mean temperature difference, °C
$\rho_g$ = Density of gas, $Kg/m^3$
$\rho_s$ = Density of shell, $ton/m^3$
$\mu$ = Viscosity of gas, kg/mh
$V$ = Gas velocity, m/s
$W_g$ = Gas flow, kg/h
$W_s$ = Weight of shell, tons
$\eta$ = Efficiency of fan
$\Delta P_g$ = Gas pressure drop, mm water column

and $h_{io}$ from the equation:

$$h_{io} = h_i d_i / d \tag{4}$$

If $V$ is the velocity of gas to be used inside the tubes,

$$V = 4W_g / (3600 \rho_g \pi d_i^2 N) \tag{5}$$

$$\rho_g = 273 MW / (22.4(273 + t_g)) \tag{6}$$

Assuming a triangular pitch arrangement of 1.25 times the diameter, the shell diameter, $D_s$, may be approximated as:

$$D_s = 1.36 d \sqrt{N} \tag{7}$$

Using an allowable stress of 1230 kg/cm² (from ASME Sec. I–Stress tables for materials), we have, for the pressure of 25 kg/cm² (a):

$$t_s = 0.01 D_s$$

Hence the weight of the shell,

$$W_s = \pi D_s t_s L_s \rho_s \tag{8}$$

$$\text{Cost, } C_s = W_s(8)\ 5000 = 40{,}000\, W_s \tag{9}$$

$$\text{Cost of tubes, } C_t = 150A \tag{10}$$

Gas pressure drop inside the tubes, allowing for 1.5 times the velocity head for entrance and exit loss is:

$$\Delta P_g = \left[ \frac{1.5 V^2 \rho_g}{2g} + \frac{f L_s V^2 \rho_g}{2g\, d_i} \right] \tag{11}$$

Substituting for $\rho_g$ and $f$ (0.023, an average value)

$$\Delta P_g = 0.0283\,[1 + (0.0153 L_s / d_i)]\, V^2 \tag{12}$$

Additional fan power required to move the gases through the tubes is given by:

$$kw = W_g \Delta P_g / (102 \eta \rho_g 3600) \tag{13}$$

Assuming 8000 hours of operation per annum and electricity cost at 30 mills/kwh

$$C_o = kw\,(8000)\,0.03F \tag{14}$$

$F$, the cost of capitalization factor is given by the formula:

$$F = \left(\frac{1+e}{1+i}\right) \left[ \frac{1 - \left(\frac{1+e}{1+i}\right)^{N_y}}{1 - \left(\frac{1+e}{1+i}\right)} \right] \tag{15}$$

$$\text{Total capitalized cost} = C_o + C_s + C_t \tag{16}$$

Hence using Equations (1) to (16), $A$, $\Delta P_g$ and $C_c$ were computed using the three tube sizes available with the help of a calculator for various velocities.

Results are presented in Figs. 2 and 3.

## *Conclusions*

1. The smallest tube size, i.e., $\phi 31.8 \times 2.9$ gives the most economical alternative.
2. Optimum gas velocity lies around 20 m/s.

Hence, detailed design and projecting can be carried out based on these inputs.

Using the procedure outlined, engineers may arrive at the optimum choice in their own problem situation.

## *References*


1. Ganapathy, V. et al., "Physical properties of selected gas streams," *Chemical Engineering,* Feb. 28, 1977.
2. Ganapathy, V., "Quick estimation of gas heat-transfer coefficients," *Chemical Engineering,* Sept. 13, 1976.

# 16. Evaluating motors using life cycle costing —

PLANT ENGINEERS SHOULD KNOW THE BASIS for selecting a motor for a given service when several alternatives are available and all appear equally good. By investigating the economics of owning and operating a motor, the best choice can be arrived at. The technique is known as life-cycle costing.

Life-cycle costing computes the present worth of all operating costs to be spent over the life of the motor. If $C_i$ is the operating cost in a year $i$, the present worth of the expenditure in the year $i$ is:

$$PV_i = C_i \left(\frac{1}{1+i}\right)^i \qquad (1)$$

The total present worth $PV_T$, considering a life of $T$ years, is:

$$PV_T = C_1\left(\frac{1}{1+i}\right) + C_2\left(\frac{1}{1+i}\right)^2 + C_T\left(\frac{1}{1+i}\right)^T \qquad (2)$$

Where $C_2$, $C_1$ etc. are the operating costs in the years 1, 2, etc. If the operating cost $C_i$ is assumed not to vary much during the life of the motor and its effect can be taken care of by using an escalation factor $e$, then:

$$PV_T = C_a\left[\left(\frac{1+e}{1+i}\right) + \left(\frac{1+e}{1+i}\right)^2 + \left(\frac{1+e}{1+i}\right)^T\right] \qquad (3)$$

Where $C_1 = C_2 = C_T = C_a$ = annual operating cost. Simplifying, we have:

$$PV_T = C_a\left(\frac{1+e}{1+i}\right)\left[\frac{1-\left(\frac{1+e}{1+i}\right)^T}{1-\left(\frac{1+e}{1+i}\right)}\right] = C_aF \qquad (4)$$

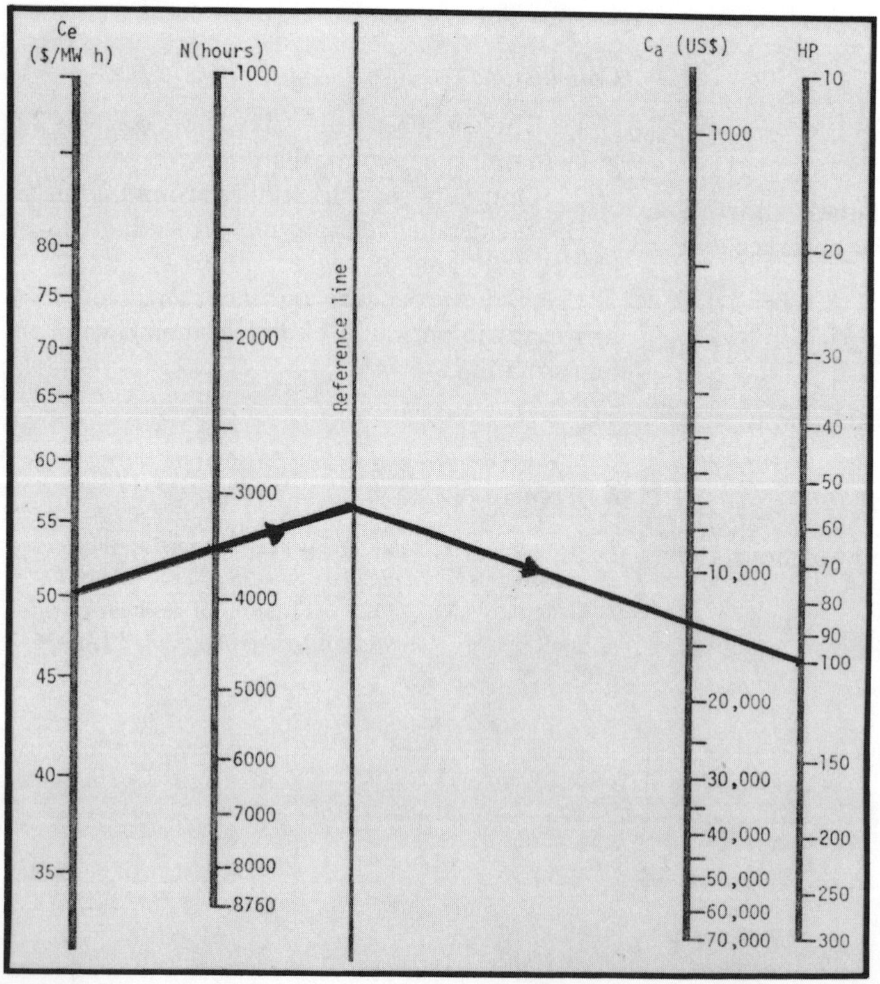


Fig. 1. Annual operating costs at 100% efficiency

***Nomenclature***

| | | |
|---|---|---|
| $C_a$ | = | Annual cost of operation ($) |
| $C_e$ | = | Cost of electricity, $/MWh |
| $e$ | = | Escalation rate, fraction |
| $i$ | = | Interest rate, fraction |
| $FC$ | = | Fixed or initial cost ($) |
| $CC$ | = | Capitalized cost ($) |
| $PV_T$ | = | Total present worth ($) |
| $T$ | = | Life of the motor (years) |
| $N$ | = | Period of operation in a year (hours) |
| $\eta$ | = | Efficiency of the motor, fraction |
| HP | = | Motor horsepower |

If the capital cost is $FC$, the total capitalized cost $CC$ is given by:

$$CC = FC + PV_T = FC + C_a F \tag{5}$$

The motor with the least $CC$ is then considered the best, as it gives the lowest cost of owning and operating the motor over its anticipated life.

The annual cost of operation $C_a$ is given by:

$$C_a = 0.746 \times \text{HP} \times N \times C_e / (1000\eta) \tag{6}$$

Fig. 1 gives $C_a$ assuming 100% efficiency and Fig. 2 gives $F$, the factor involving interest rate, escalation and life.

### *Example*

Motors are quoted by two different vendors (see table above right for details). Which motor gives the lowest life-cycle cost?

| | A | B |
|---|---|---|
| Motor capacity, HP | 100 | 100 |
| Cost, $ | 4200 | 5800 |
| Efficiency, % | 88 | 92 |
| Life, years | 20 | 20 |
| Cost of electricity | 50 | 50 |
| Period of operation in a year | 3500 | 3500 |
| Interest rate, % | 12 | 12 |
| Escalation, % | 7.5 | 7.5 |

### *Solution*

In Fig. 1, connect $C_e = 50$ with $N = 3500$ and extend to cut line 1. Connect with HP = 100 to cut $C$ scale at $13,000.

Actual $C_a$ of motor A, considering $\eta$ is (13,000/0.88) = $14,770.

That of motor B is (13,000/0.92) = $14,130.

In Fig. 2, go up from $i = 12\%$ to cut $e = 7.5\%$. Move horizontally to cut $T = 20$. Move down to cut $F$ scale at 11.3.

Hence $CC$ of motor A is (4200 + 11.3 × 14,770) = $171,101.

That of motor B is (5800 + 11.3 × 14,130) = $165,469.

Thus, although the initial cost of motor B was more, in the long run it works out cheaper.

Let us see what happens if we use the motor for only 500 hours in a year. In this case:

$C_a$ for motor A is (500/3500) × 13,000 = $1875 (100% efficiency).

$CC$ for motor A is (4200 + 11.3 × 1857/0.88) = $28,046.

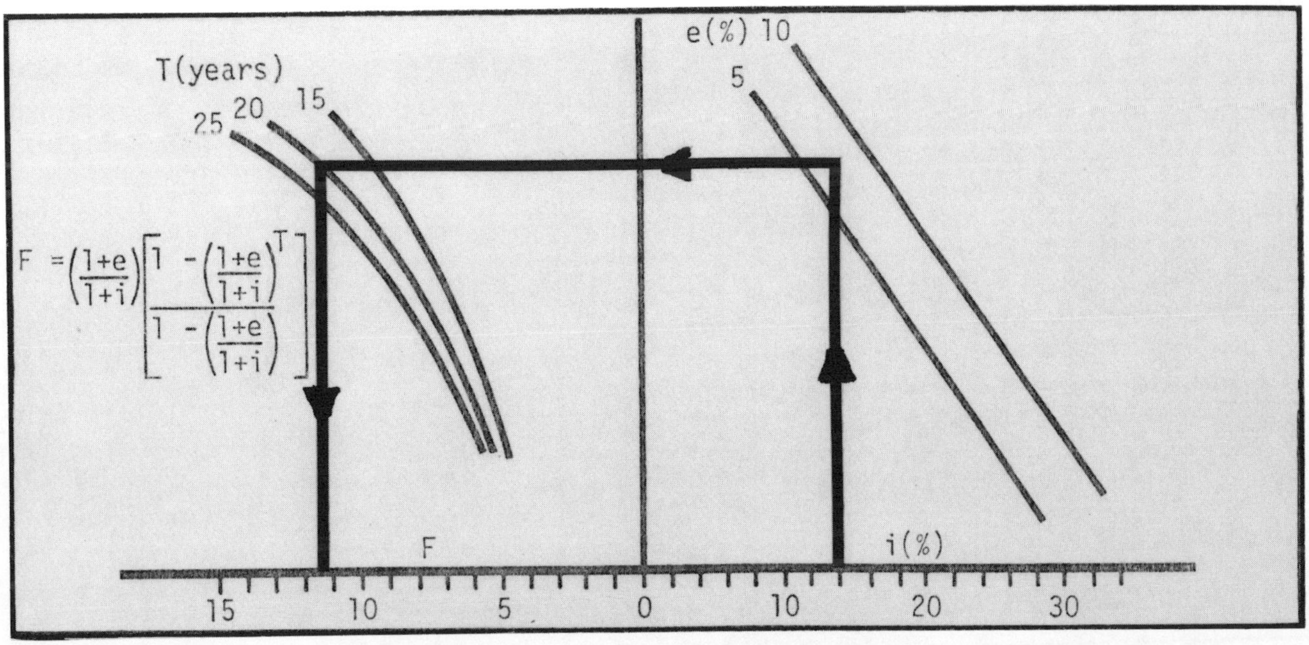


**Fig. 2. Derivation of factor *F*, given *i*, *e* and *T***

*CC* for motor B is (5800+11.3×1857/0.92= $28,609.

We now see that motor B has a higher life cycle cost, so motor A is a better choice when the period of operation is low. Plant engineers can thus work out *CC* for various *N, e, i, T* values and choose the motor that gives the lowest *CC*. This study also shows that when the period of operation is low, a high cost motor is not justified. Note: cost of maintenance can also be considered by multiplying $C_a$ by a certain factor, based on experience.

# Chapter 5

# Fluid Flow and Pressure Drop Calculations

# 1. Nomogram converts pressure and flow units for liquids

IN MANY PUMPING APPLICATIONS, THE PRESSURE to be developed in pounds per square inch and the flow in pounds per hour will be known. However, for pump selection, the head in feet of liquid and the flow in gallons per minute are desirable. The nomogram simplifies the conversion. It is only necessary to know the specific gravity or density of the liquid.

Equations used are:

$$Q=0.124\ W/\rho$$

$$H=144\ P/\rho$$

Where:

$Q$ = Volume, gpm
$\rho$ = Density, lb/ft$^3$
$W$ = Mass flow, lb/hr
$H$ = Head, ft of liquid
$P$ = Pressure, psi

### *Example*

50,000 lb/hr of a liquid with a specific gravity of 0.9 has to be raised through a pressure of 400 psi. Find volume flow in gpm and head in feet of liquid.

### *Solution*

Connect gravity of 0.9 with pressure of 400 and extend to cut the head scale. Read answer as 1020 ft. Connect specific gravity of 0.9 with mass flow of 50,000 and extend to cut the $Q$-scale. Read answer as 110 gpm.

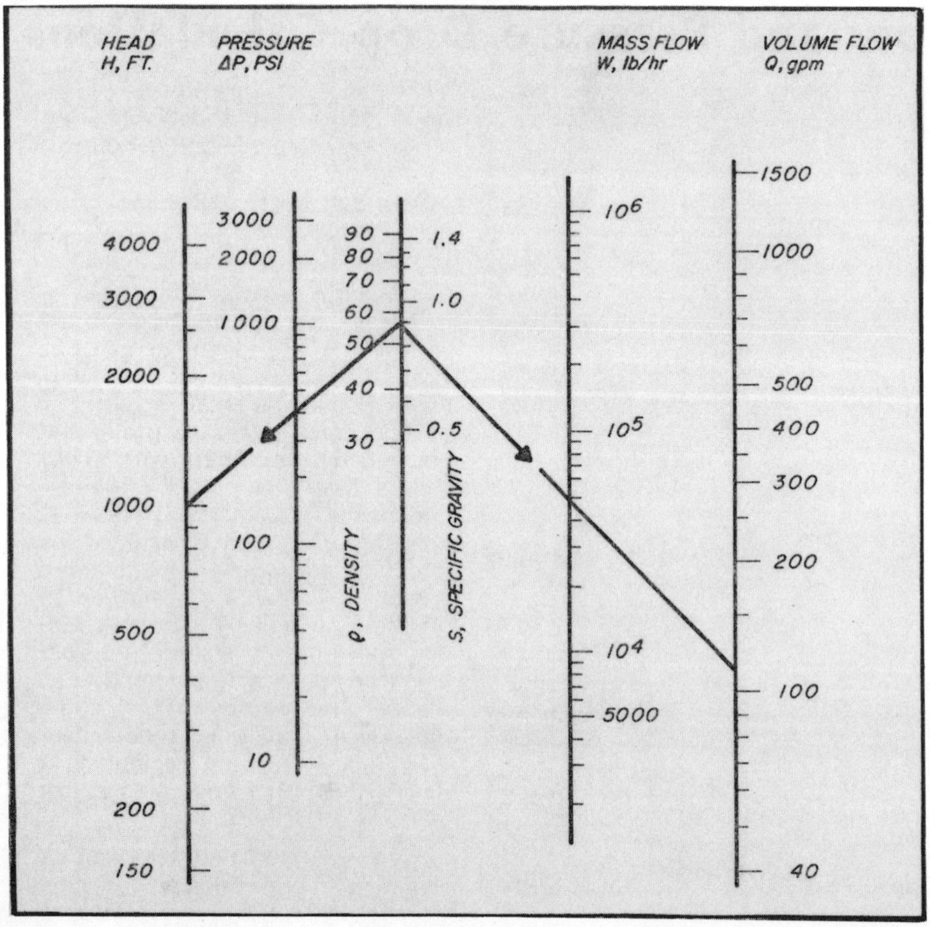

# 2. Nomogram relates mass and volumetric flow of gases

IN DEALING WITH THE FLOW of gases and compressed air, it is often necessary to convert between mass flow and volumetric flow. The relationship is comparatively simple, but it involves finding the density of each gas at different pressures and temperatures. This nomogram makes tables unnecessary.

Density is found first from the relationship:

$$\rho = \frac{144\,PM}{1545\,T}$$

where:

$\rho$ = density, lb/cu ft
$P$ = pressure, psia
$M$ = molecular weight of gas (use key from table)
$T$ = temperature, R

Then, weight of flow is obtained:

$$W = 60\,\rho\,V$$

where:

$W$ = mass flow rate, lb/hr
$V$ = volumetric flow, cfm

**Table**

| | |
|---|---|
| A=hydrogen | L=argon |
| B=helium | M=carbon dioxide |
| C=methane | N=methyl chloride |
| D=ammonia | O=butane |
| E=natural gas | P=sulfur dioxide |
| F=acetylene | Q=chlorine |
| G=nitrogen and carbon monoxide | R=benzene |
| H=air | S=hexane |
| I =oxygen | T=toluene |
| J =hydrogen sulfide | U=N-heptane |
| K=hydrochloric acid | V=octane |
| | W=Freon |
| | X=Dowtherm E |

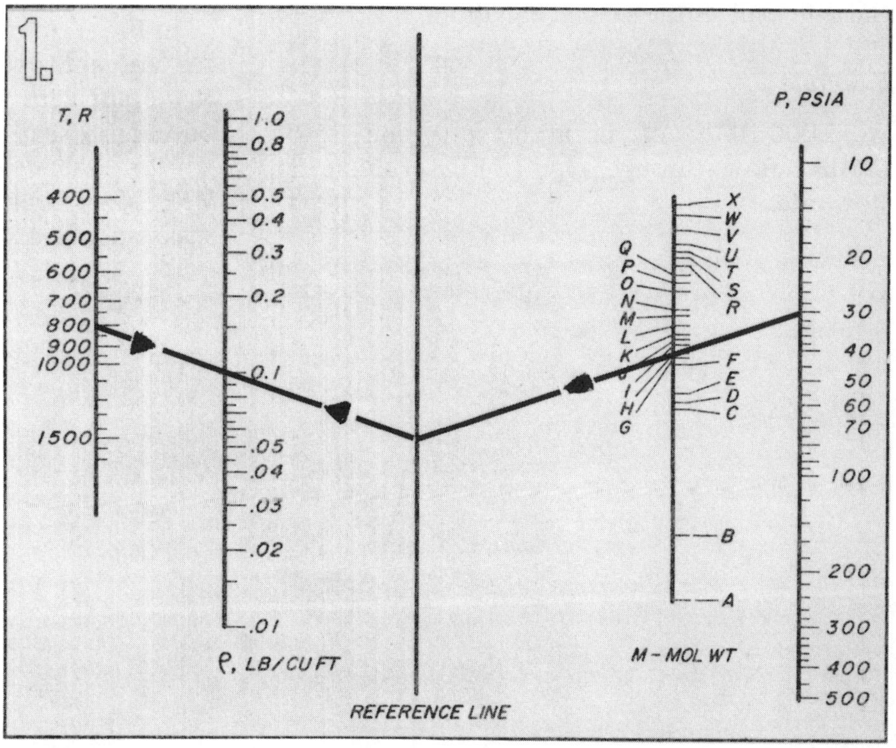


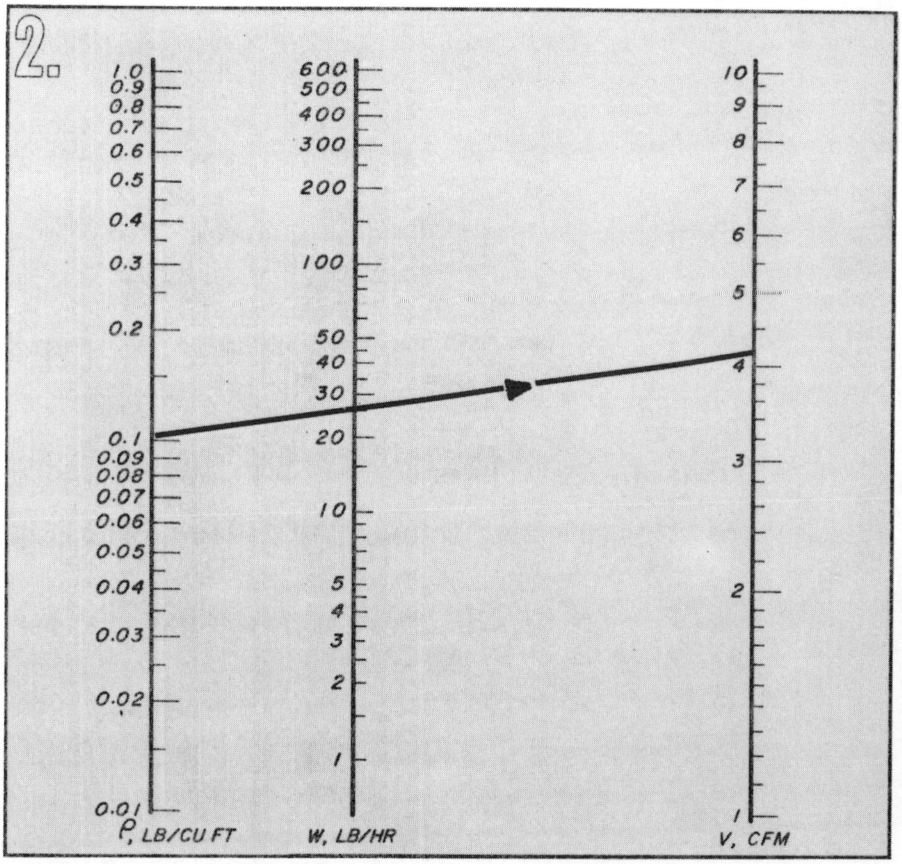

One nomogram solves the first equation, and the other solves the second equation.

### *Example*

25,000 lb/hr of air at 30 psia and 350F (810R) flow through piping. Find volumetric flow rate.

### *Solution*

In Fig. 1, join $P=30$ with point $H$ (from table, for air) and extend to cut reference line. Join this point with $T=810$ and read $\rho=0.1$ lb/cu ft on density scale. In Fig. 2, connect $\rho=0.1$ with $W=25$, extend to volume scale and read $V=4.2$. With same multiplication factor as for weight scale, answer is 4200 cfm.

# 3. Chart determines velocity of steam in pipes —

QUICK ESTIMATES OF STEAM VELOCITY IN PIPES and tubes can be made with the nomograph on the following page without reference to steam tables, thereby saving time. It is sufficient to know steam pressure and temperature to use this nomograph. It is based on the formula:

$$V = 3.06\, W_s v / d^2$$

where

$V$ = line velocity, fpm
$W_s$ = steam flow, lb per hr
$v$ = specific volume of steam, cu ft per lb
$d$ = tube or pipe inside diameter, in.

*Example*

What is the velocity of superheated steam at 1000 psia and 900F if 6000 lb per hr of it flows through a tube with a 1.5 in. inside diameter?

*Solution*

Draw a vertical line from 900F on the T scale to 1000 psia on the P scale. From this point, draw a horizontal line to Pivot Line 1. Connect this point with 1.5 in. on the d scale and extend the line to Pivot Line 2. Connect this point with 6000 lb per hr on the $W_s$ scale and extend the line to the V scale. Read 6300 fpm as the solution.

Note that if the line velocities are restricted, you can work in reverse to obtain the tube or pipe size.

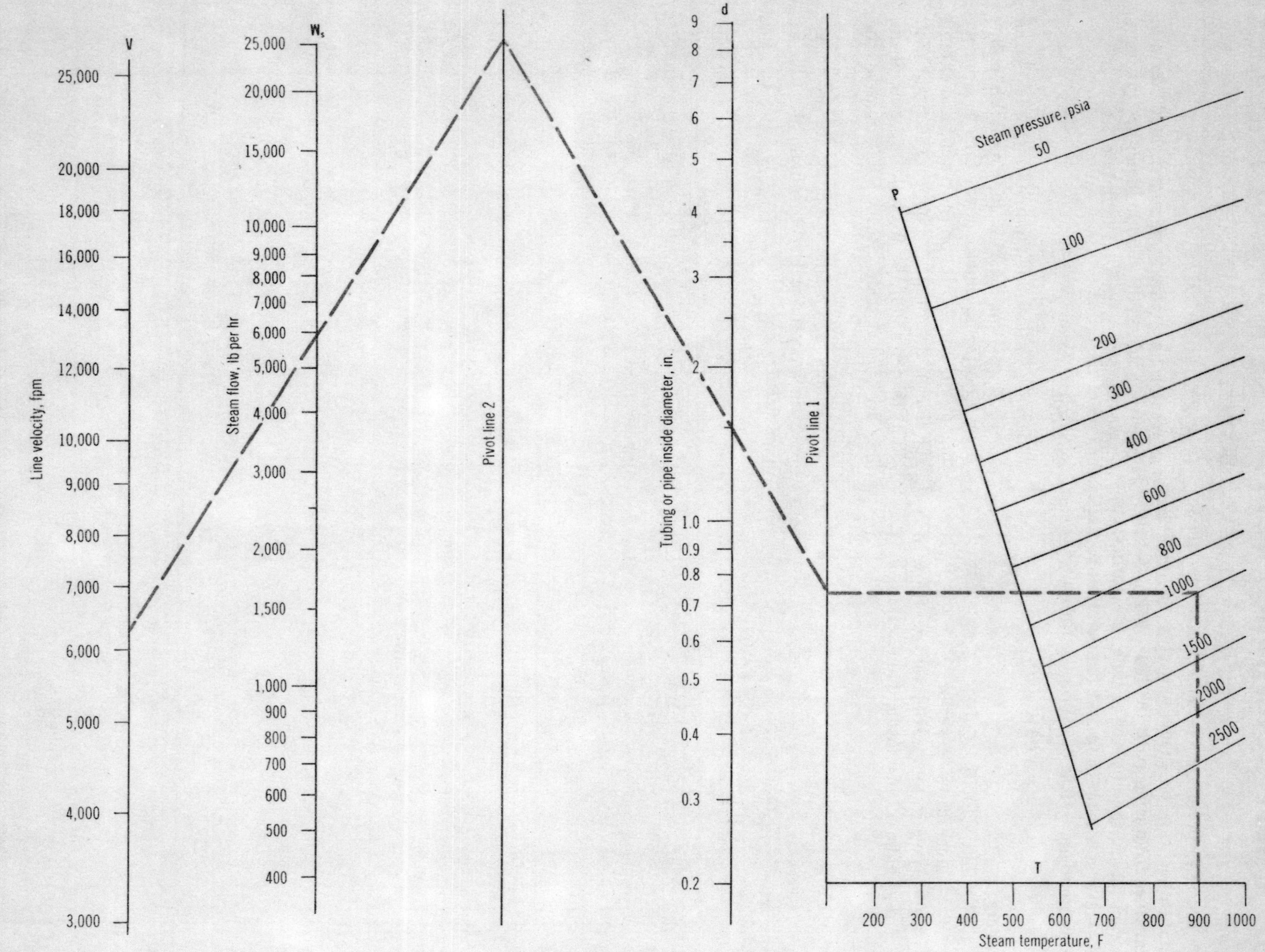


Nomograph determines steam velocity in piping. Broken lines indicate solution to example in text.

# 4. Nomogram determines Reynolds number for superheated steam

REYNOLDS NUMBER IS AN IMPORTANT QUANTITY for any fluid. It influences pressure drop and is required in heat-transfer calculations. For superheated steam, Reynolds number,

$$Re = \rho VD/\mu$$

where:

$\rho$ = density of superheated steam
$D$ = inside diameter of pipe or tube
$\mu$ = absolute viscosity

All variables must be in consistent units.

Converting the above equation by introducing mass flow rate gives:

$$Re = 15.2\, G/D\mu$$

where:

$G$ = mass flow of superheated steam, lb/hr
$D$ = inside diameter of tube or pipe through which steam flows, in.
$\mu$ = absolute viscosity, lb/ft-hr

The absolute viscosity of superheated steam does not vary as much with the steam pressure as with steam temperature. For practical purposes, it may be taken to depend on steam temperature. Hence, with the nomogram, it is not necessary to refer to tables giving viscosity of superheated steam. The nomogram takes care of it.

### *Example*

What Reynolds number is obtained when $5 \times 10^5$ lb/hr of superheated steam flows through a pipe with an inner diameter of 10 in.? Pressure

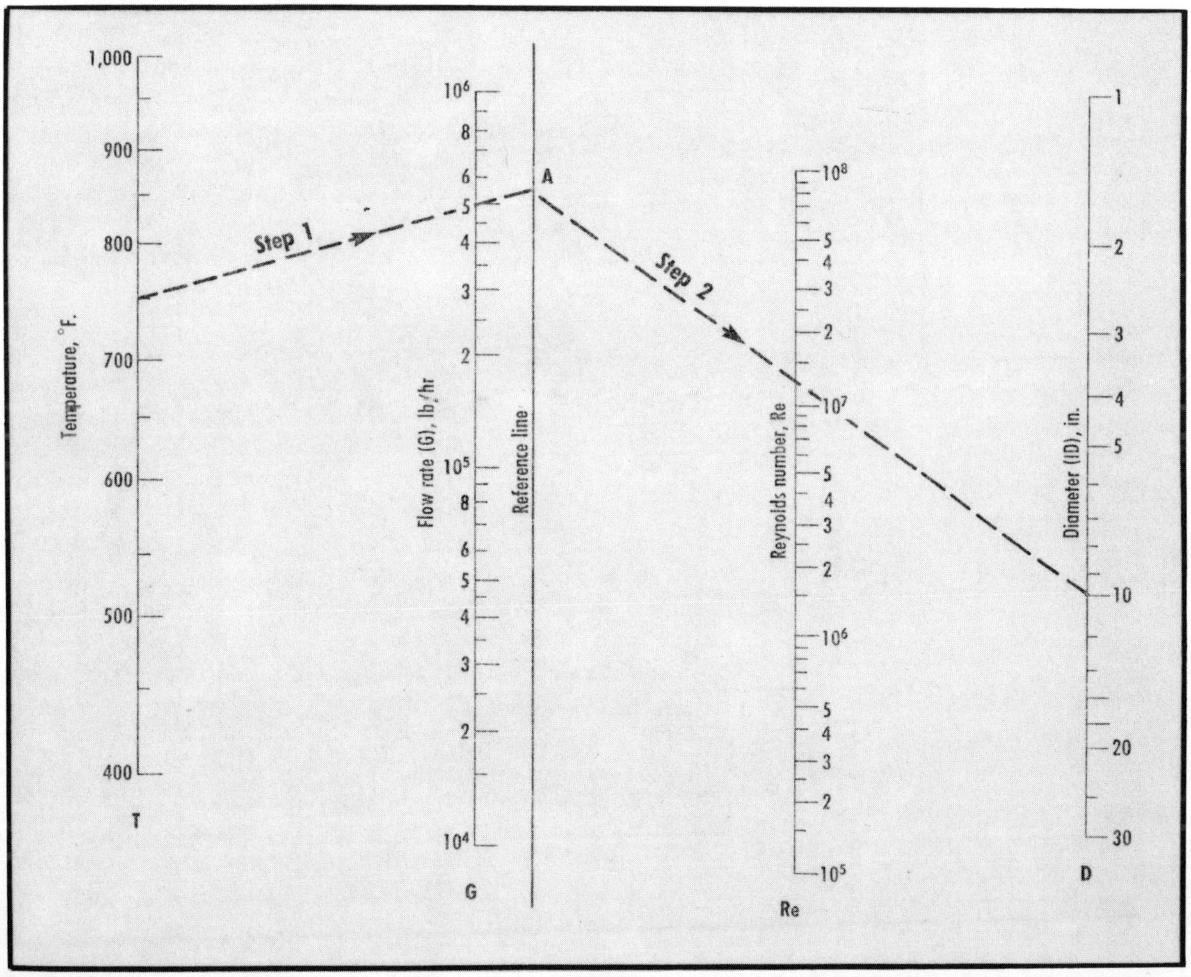

of the steam is 1,600 psig and its temperature is 750°F.

***Solution***

Join 750 on the T scale (Step 1) with $5 \times 10^5$ on the G scale and extend to cut the reference line at A. Next (Step 2) join A with 10 on the D scale. Read Reynolds number Re as $1.26 \times 10^7$.

Since Reynolds number (Re) varies directly with flow rate (G), the same multiplication factor may be used for both. For example, if G is $5 \times 10^3$ lb/hr, Re is $1.26 \times 10^5$ where $10^2$ is the multiplication factor.

# 5. Nomogram gives pressure loss in steam lines

IN PROCESS PLANTS, IT IS FREQUENTLY DESIRable to estimate the pressure drops in steam lines. The formula used is not difficult to solve, but reference to steam specific volume at various pressures and temperatures makes it a tedious calculation.

The alignment chart presented here provides a quick way to obtain pressure losses. It is sufficient, if pressure and temperature of the steam are known. There is no need to refer to steam tables, and velocity need not be computed.

The friction factor for turbulent flow has been considered, and the variation with diameter is taken care of by the chart. The chart is based on the formula,

$$\Delta P = 3.36 f G^2 L V / D^5 \times 10^6$$

Where:

$\Delta P$ = pressure loss, psi
$f$ = friction factor, a function of diameter
$G$ = mass flow, lb/hr
$V$ = specific volume, cu ft/lb
$L$ = total equivalent length, ft
$D$ = ID, in.

### *Example*

Assume 2,000 lb/hr of superheated steam flows through a line with an ID of 2.0 in. Steam pressure is 800 psi, and steam temperature is 700°F. Equivalent length of pipe is 200 ft.

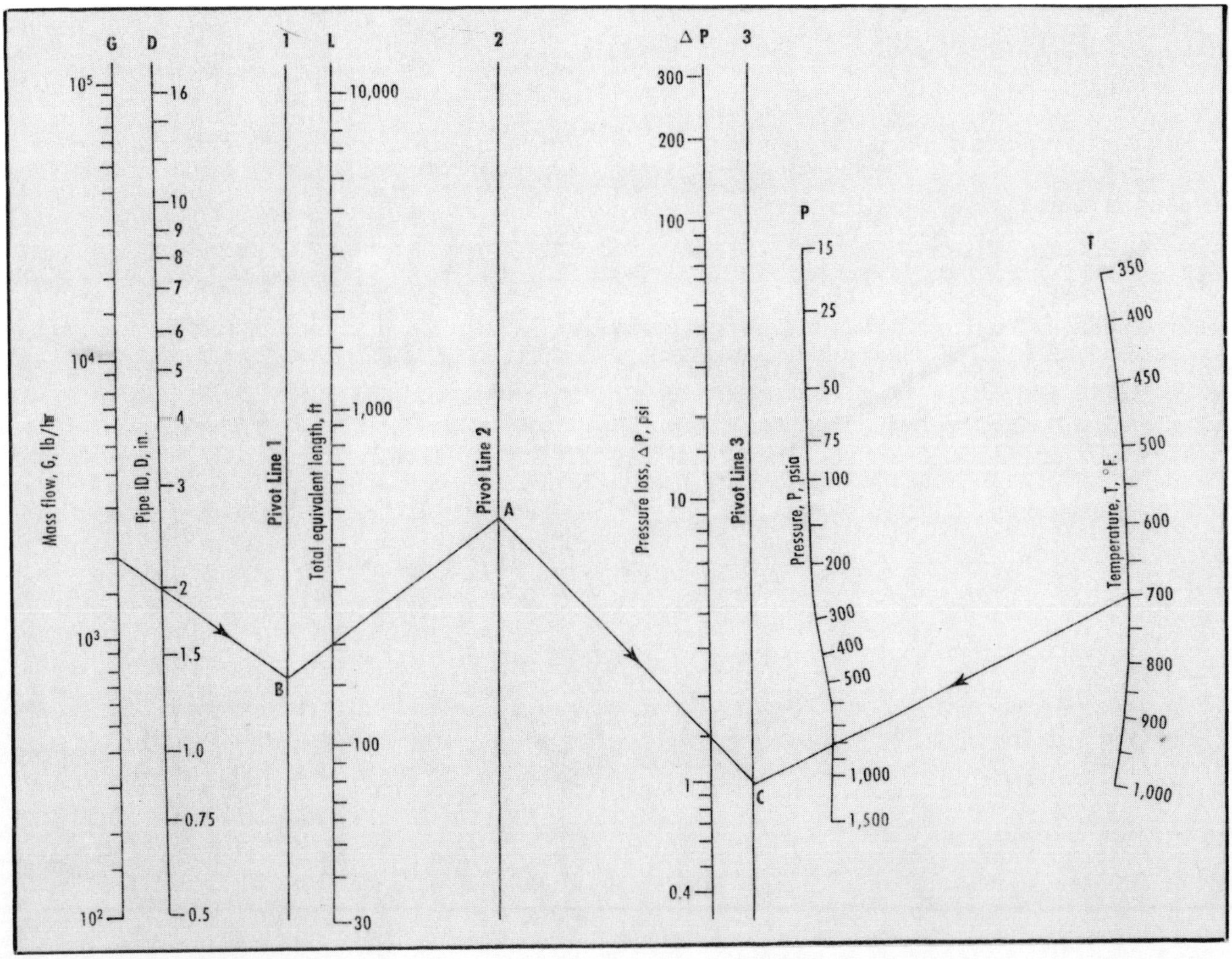

What is the pressure drop through the line?

***Solution***

On the alignment chart, connect 700°F on the T scale with 800 psi on the P scale, and mark the intersection on Pivot Line 3 at C. Then connect 2,000 on the G scale with 2 on the D scale and extend to Pivot Line 1 at Point B. Next connect Point B with 200 on the L scale and extend to intersect Pivot Line 2 at Point A. Join Point A with Point C and read $\Delta P = 1.5$ psi.

# 6. Chart relates steam mass flow rate to velocity

ENGINEERS INVOLVED IN SELECTION OF SUPERheaters or reheaters go by mass flow rates.

But it is desirable to have an idea of the velocity. In sizing piping or tubes, one would like to convert mass flows to velocities to see if they do not exceed the recommended velocities.

In these situations, reference to steam tables to obtain specific volume *(v)* in cu ft/lb is a must which means a considerable waste of time.

This chart converts mass flows $(w_s)$ in lb/hr to mass flow rates *(m)* in lb/ft²-hr and velocity *(V)* in ft/sec without the aid of steam tables. Pipe ID *(d)* in inches must be known.

The recommended steam mass flow rates in boiler or fired superheaters vary from 200,000 to 700,000 lb/ft²-hr, while velocities in pipes vary from 150 to 300 fps.

***Example***

If 6,000 lb/hr of superheated steam at 1,000 psia and 800°F flows through a pipe with an ID of 1.5 in., what is the mass flow rate and velocity?

The chart solves the following equations:

$$m = 183\, w_s/d^2 \qquad (1)$$

$$V = mv/3{,}600 \qquad (2)$$

***Solution***

Connect $w_s = 6{,}000$ with $d = 1.5$ in. to cut scale m at 490,000 lb/ft²-hr. On the right-hand side of the chart, go vertically from T=800 to P=1,000 and move left to cut line 1 at 0. Connect that point with m=490,000 and read V= 90 fps.

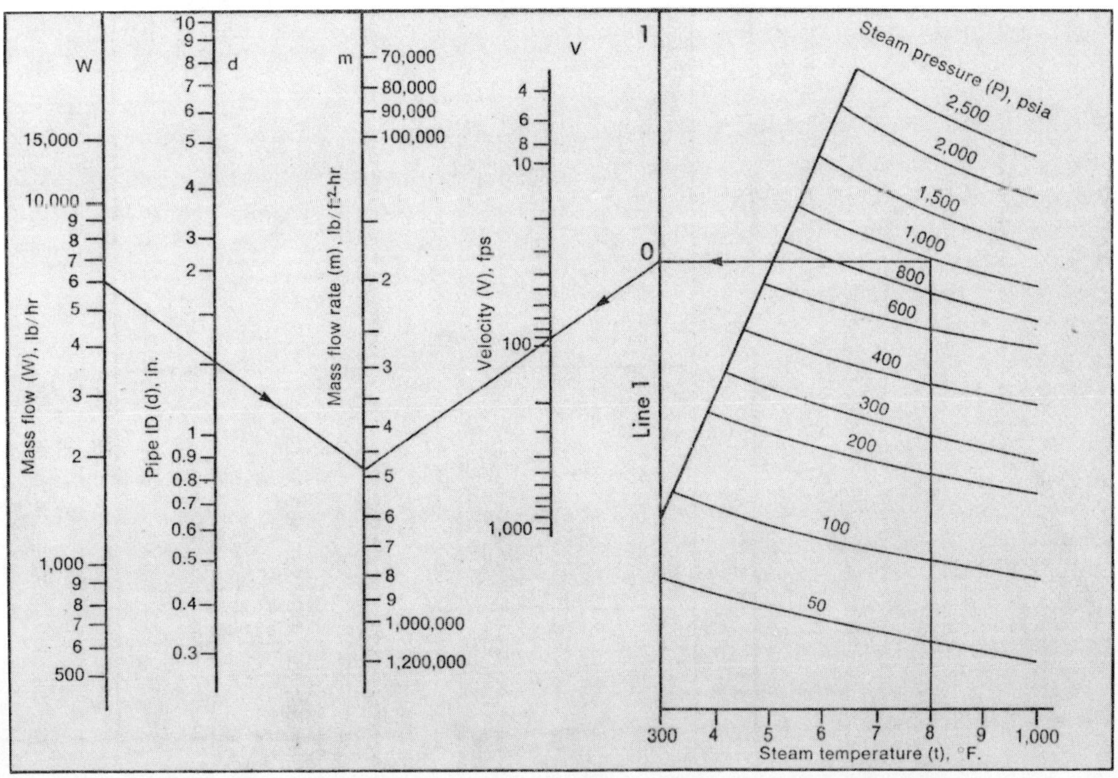

# 7. Chart estimates pressure drop in smooth pipes and tubes

WATER FLOWING THROUGH A SMOOTH TUBE OR pipe is subject to pressure loss in the direction of flow. The amount of pressure loss can be estimated using the equation

$$\Delta p = \frac{f L_e V^2 \rho}{24 g d_i} \qquad (1)$$

where:

- $\Delta p$ = pressure drop per 100 ft of length, psi
- $f$ = friction factor, dimensionless
- $L_e$ = equivalent length of pipe or tube, ft
- $V$ = velocity of water, fps
- $\rho$ = density of water, lb/cu ft
- $g$ = acceleration due to gravity, 32 ft/sec$^2$
- $d_i$ = tube or pipe inside diameter, in.

Friction factor $f$ is a function of the Reynolds number:

$$f = 0.133\,\mathrm{R}^{-0.174} \qquad (2)$$

where:

- R = Reynolds number, dimensionless

Defining the Reynolds number in Equation 2, as

$$\mathrm{R} = \frac{\rho V d_i}{12\mu}$$

where:

- $\mu$ = viscosity of water, lb-sec/sq ft

and combining with, and simplifying, Equation 1, gives:

$$\frac{\Delta p}{L_e} = \frac{0.000267\,\rho^{0.826}\,\mu^{0.174}\,V^{1.826}}{d_i^{1.174}} \qquad (3)$$

Because $\rho$ and $\mu$ are functions of temperature, a further simplification can be made by precalculating the term $\rho^{0.826}\,\mu^{0.174}$. This term has been precalculated in the preparation of this

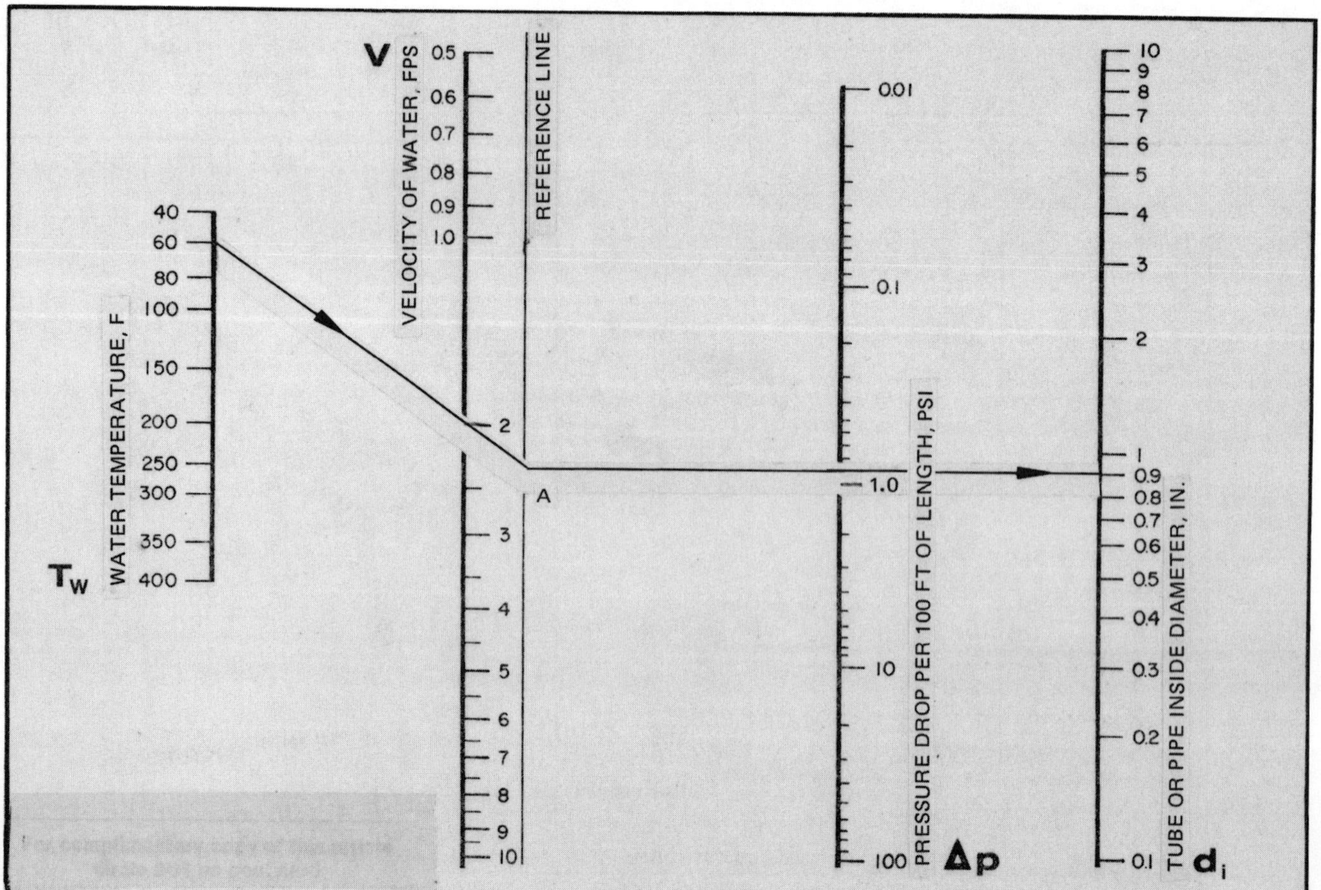

nomograph; thus, a specific problem can be solved without reference to tables for viscosity or to Moody's chart for friction factor values. The nomograph gives the pressure drop per 100 ft as a function of temperature, velocity, and tube size.

### *Problem*

Determine the pressure drop per 100 ft of length when 60F water flows through a 0.9 in. ID tube at a velocity of 2 fps.

### *Solution*

Connect 60F on Scale $T_w$ with 2 fps on Scale V and extend the line to intersect line 1 at point A. Connect point A with 0.9 in. on Scale $d_i$, and, where Scale $\Delta p$ is crossed, read pressure drop as 0.92 psi.

# 8. Charts evaluate gas-pressure drop in tube banks, plain and finned

TUBE BUNDLES USED IN BOILER superheaters, economizers, fired heaters, or process heaters offer resistance to the flow of flue gases or air flowing over them. Depending on the tube geometry, pitch arrangement, temperature, and gas velocity used, the gas-pressure drop will vary.

Gas-pressure drop can be costly in the long run as fan power consumption converted over the life of the equipment can be substantial. For example, if the additional fan power is 10 kw because of using high gas velocities, the capitalized cost could be about \$20,000, depending on cost of electricity, escalation, and interest rates.

Several correlations are available in the literature to predict the gas-pressure drop. Each organization develops its own data based on site operation and experience, which should be used to give accurate results.

In the absence of such data, the gas-pressure drop over a bundle of plain tubes may be predicted as follows for estimation purposes.

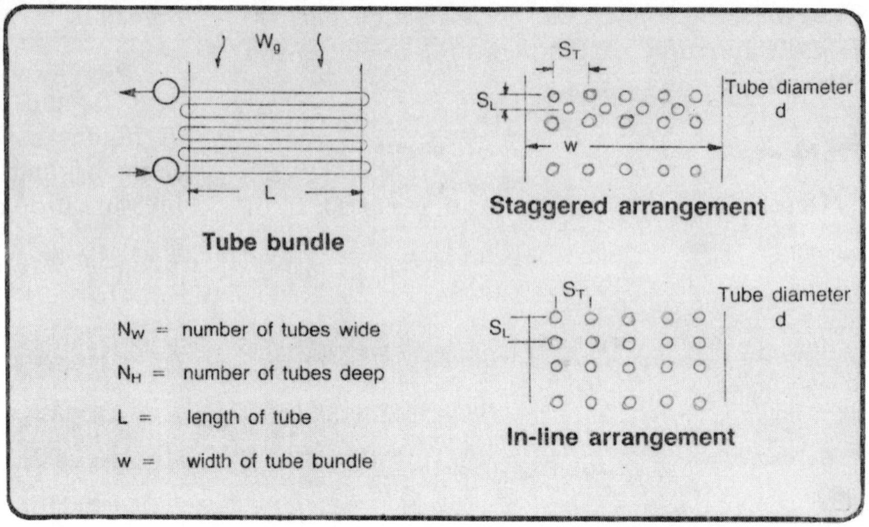


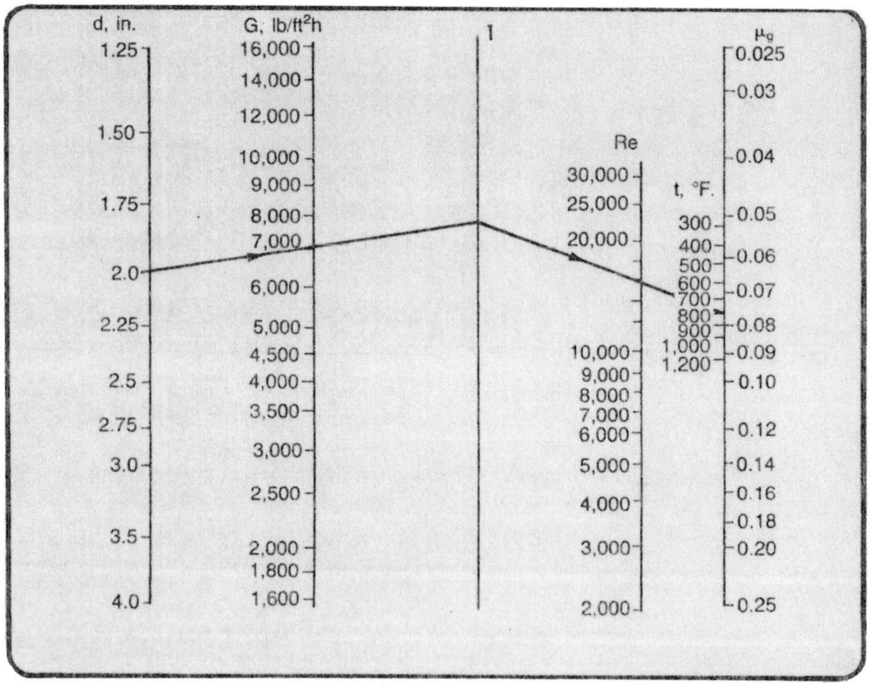


**Fig. 2. To find Reynolds Number**

### *Plain Tubes*

$$\Delta P_g = 9.3 \times 10^{-10} f_g G^2 N_H / \rho_g \qquad (1)$$

Where the friction factor, $f_g$, is given for $2000 < Re < 40{,}000$ by the following correlations.[1]

For an in-line arrangement:

$$f_g = Re^{-0.15} \left[ 0.044 + \frac{0.08\,(S_T/d)}{(S_T/d - 1)^{(0.43 + 1.13 d/S_L)}} \right] \qquad (2)$$

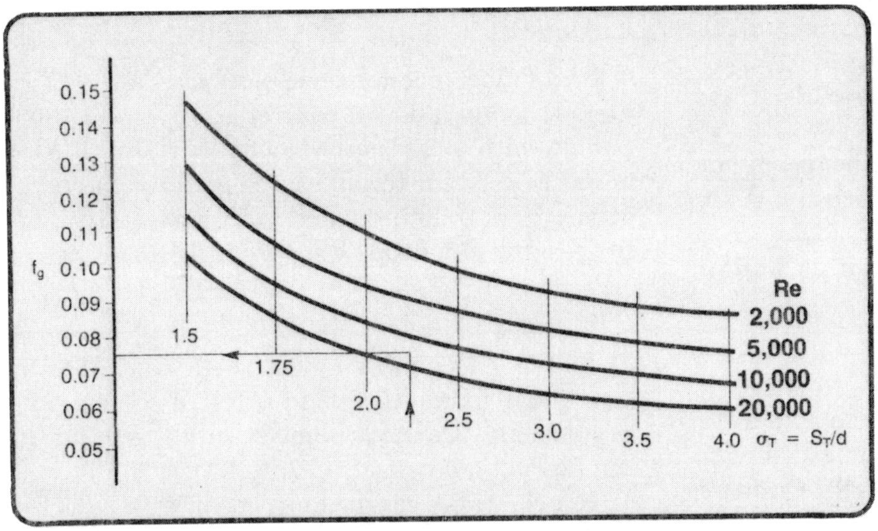


Fig. 3. Friction factor for staggered arrangement

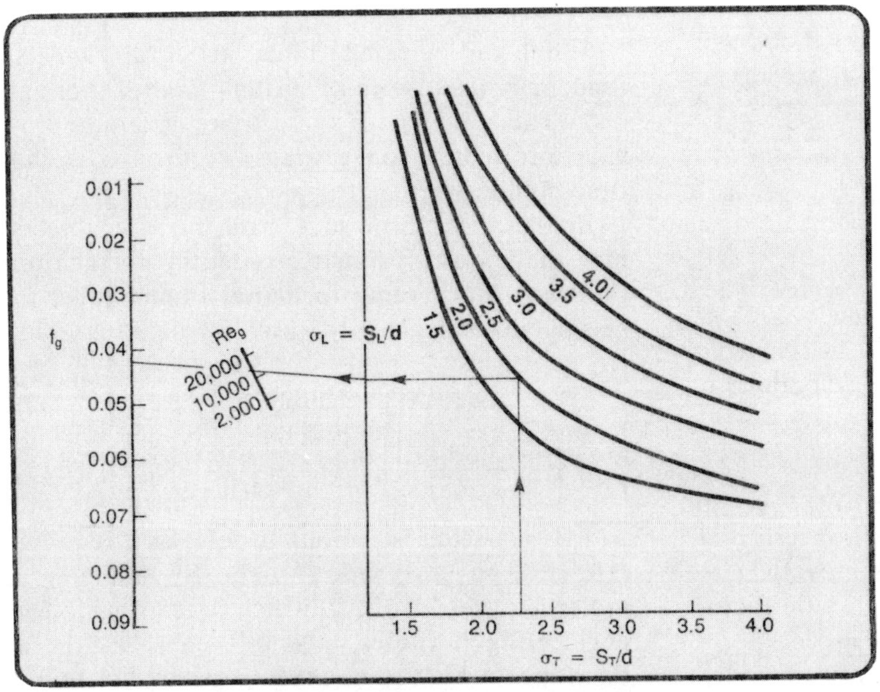


Fig. 4. Friction factor for in-line arrangement

For a staggered arrangement:

$$f_g = \mathrm{Re}^{-0.16}\left[0.25 + \frac{0.1175}{(S_T/d - 1)^{1.08}}\right] \quad (3)$$

Re, the gas Reynolds Number, is given by:

$$\mathrm{Re} = \frac{Gd}{12\mu} \quad (4)$$

### Nomenclature

| | |
|---|---|
| $A_o$ | Obstruction area of finned tube, ft$^2$/ft |
| $b$ | Fin thickness, in. |
| $d$ | Tube outer diameter, in. |
| $f_g$ | Friction factor |
| $G$ | Gas mass velocity, lb/ft$^2$h |
| $h$ | Fin height, in. |
| $L$ | Length of tube bundle, ft |
| $MW$ | Gas molecular weight |
| $n$ | Number of fins/in. |
| $\Delta P_g$ | Gas pressure drop, in. WC |
| Re | Reynolds number |
| $\rho_g$ | Density of gas, lb/cu ft |
| $S_T$, $S_L$ | Transverse and longitudinal pitch of tubes, in. |
| $t$ | Gas temperature, °F |
| $V_g$ | Gas velocity, ft/sec |
| $w$ | Width of bundle, ft |
| $W_g$ | Gas flow, lb/h |
| $\mu$ | Gas viscosity, lb/ft h |
| $\sigma_T$, $\sigma_L$ | Ratios $(S_T/d)$ and $(S_L/d)$, respectively |

Where $G$ is the gas mass velocity. The maximum value of $G$ is to be used in the above equations.

This maximum occurs in the plane transverse to the gas flow.

$$G = \frac{12\,W_g}{L[w - N_w(S_T - d)]} \quad (5)$$

Fig. 1 is a typical arrangement of tubes. Though the equations are not particularly difficult, more time is spent on obtaining the viscosity and density of flue gases in order to estimate the Reynolds Number and pressure drop.

Fig. 2 has been developed to give Re if gas mass velocity $G$, tube diameter $d$, and flue gas or air temperature are known. One need not look up gas property tables for viscosity. Also, if viscosity is known, Re may be found. The chart is reasonably accurate for air as well as flue gases.

### *Example 1*

Flue gases flow over a tube bundle with a gas mass velocity of 7,000 lb/ft$^2$h. Tube diameter is 2.0 in. and gas temperature is 800°F. Estimate the Reynolds Number.

### *Solution*

On Fig. 2, connect $d=$ 2.0 with $G=7{,}000$ and extend to cut line 1. Connect with $t=800$ to cut Re scale at 16,000.

Figs. 3 and 4 give the friction factor, $f_g$, for staggered and in-line arrangements.

### *Example 2*

If the ratio $\sigma_T=(S_T/d)=2.2$ and $\sigma_C=S_L/d=2.0$, estimate the friction factor in staggered and in-line tube bundles of a boiler evaporator.

### *Solution*

For staggered tube bundles, in Fig. 3, go up from $\sigma_T=2.2$ to cut Re$=16{,}000$ (an error of 10% in Re hardly matters), and move left to cut $f_g$ scale at 0.074. This is the friction factor for staggered arrangement; $\sigma_L$ is not used.

In Fig. 4, for in-line bundles, go up from $\sigma_T=2.2$ to cut $\sigma_L=2.0$. Move left to cut reference line. Connect with Re$=16{,}000$ and extend to cut $f_g$ scale at 0.043.

Fig. 5 gives the gas-pressure drop per row of tubes if $G$, $t$, and $f_g$ are known based on equation 1. Also, the velocity, $V_g$, is obtained based on:

$$V_g=\frac{G}{3600\,\rho_g} \tag{6}$$

There is no need to calculate the gas density $\rho_g$.

### *Example 3*

In a boiler economizer, flue gases at 800°F average temperature flow over 16 rows of a tube bundle with a gas mass velocity of 7,000 lb/ft$^2$ h. Friction factor found using Fig. 3 for the tube geometry is 0.085. Estimate the gas velocity and the gas-pressure drop.

### *Solution*

In Fig. 5, connect $t=800$ with $G=7{,}000$ to cut line 1 at A. Also, read off $V_g=63$ ft/sec as gas velocity. Connect point A with $f_g=0.085$ to cut $(\Delta P_g/N_H)$ scale at 0.12 in. WC.

Hence, total gas pressure drop $=16\times0.12=$ 1.92 in. WC.

### *Finned Tubes*

Spiral and serrated finned tubes are widely used in boilers, process heaters, and economizers.[2] The suppliers of these tubes generally provide information on gas-pressure drop over the tube bundles.

In the absence of such data, plant engineers may make quick estimates using the correlation of Briggs and Young for spiral finned tubes in staggered arrangement, which is often used:[3,4]

$$\Delta P_g=1.58\times10^{-8}\left[\frac{G^{1.684}\,d^{0.611}\,\mu^{0.316}}{S_T^{\,0.412}\,S_L^{\,0.515}}\,\frac{(460+t)}{MW}N_H\right] \tag{7}$$

This equation is tedious to solve as it requires reference to gas property data. By assuming a molecular weight of 30, which is reasonable for engineering purposes, a simple chart has been developed which gives fairly good estimates for tube geometry and pitch used in industry.

Fig. 6 solves equation 7 for flue gases and may also be used for air. It is sufficient if $G$, $d$, and $t$ are known.

$G$, gas mass velocity, is obtained for spiral-finned tubes as follows:

$$G=\frac{W_g}{L(w-N_wA_o)} \tag{8}$$

$$\text{Where: } A_o=\left[\frac{d}{12}+\frac{nbh}{6}\right] \tag{9}$$

Data on $A_o$, the obstruction area of finned tubes, is supplied by manufacturers; in the absence of this data, one may use equation 9.

### *Example 4*

In a waste-heat boiler, flue gas flows over a spiral-finned economizer tube bundle with the following arrangement.

| | | |
|---|---|---|
| Gas flow, $W_g$ | = | $1.95 \times 10^6$ lb/hr |
| Tube OD, $d$ | = | 1.75 in. |
| Fin height, $h$ | = | 0.75 in. |
| Fin thickness, $b$ | = | 0.06 in. |
| Number of fins/in., $n$ | = | 2.0 |
| Longitudinal pitch, $S_L$ | = | 4.5 in. |
| Transverse pitch, $S_T$ | = | 4.5 in. |
| Number of tubes across | = | 52 |
| Finned tube length, $L$ | = | 24 ft |
| Width of bundle, $w$ | = | 20 ft |
| Average gas temperature, $t$ | = | 800°F |
| Number of tubes deep, $N_H$ | = | 6 |

Estimate the gas-pressure drop.

### *Solution*

$$A_o = \left[\frac{1.75}{12} + \frac{2 \times 0.06 \times 0.75}{6}\right]$$

$$= 0.161 \text{ ft}^2/\text{ft}$$

$$G = \frac{1.95 \times 10^6}{(20 \times 24 - 0.161 \times 52 \times 24)}$$

$$= 7{,}000 \text{ lb/ft}^2\text{h}$$

In Fig. 6, connect $G = 7{,}000$ with $t = 800$ to cut the $\Delta P_g/N_H$ scale. On $d = 1.75$ in. scale, read off 0.30 in. WC. Total gas pressure drop $= 0.30 \times 1.8$ in. WC. One may also substitute the data in equation 7 and check the variations. This has been done, and error is less than 5%.

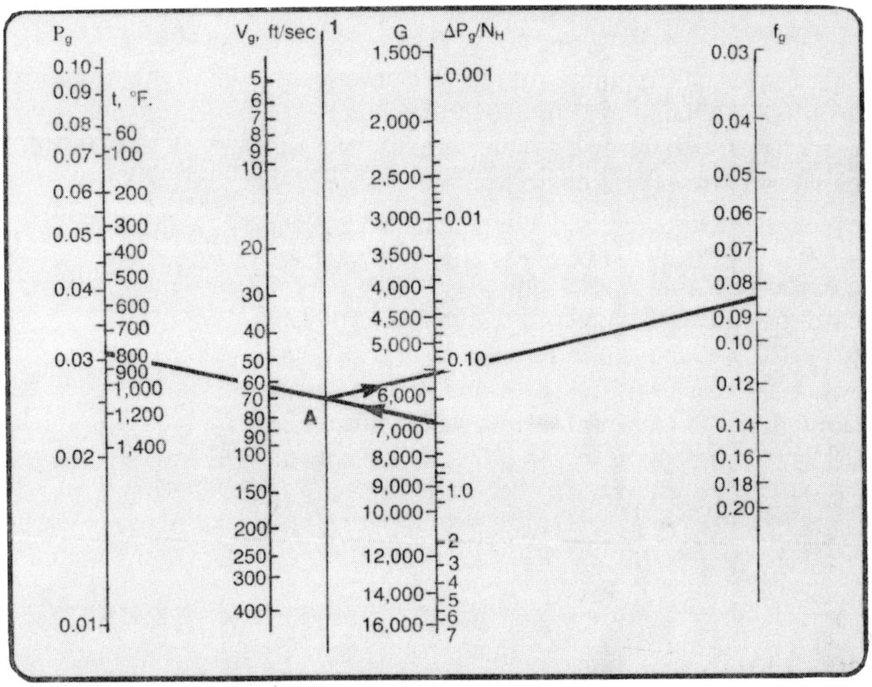


**Fig. 5. Gas velocity, pressure drop**

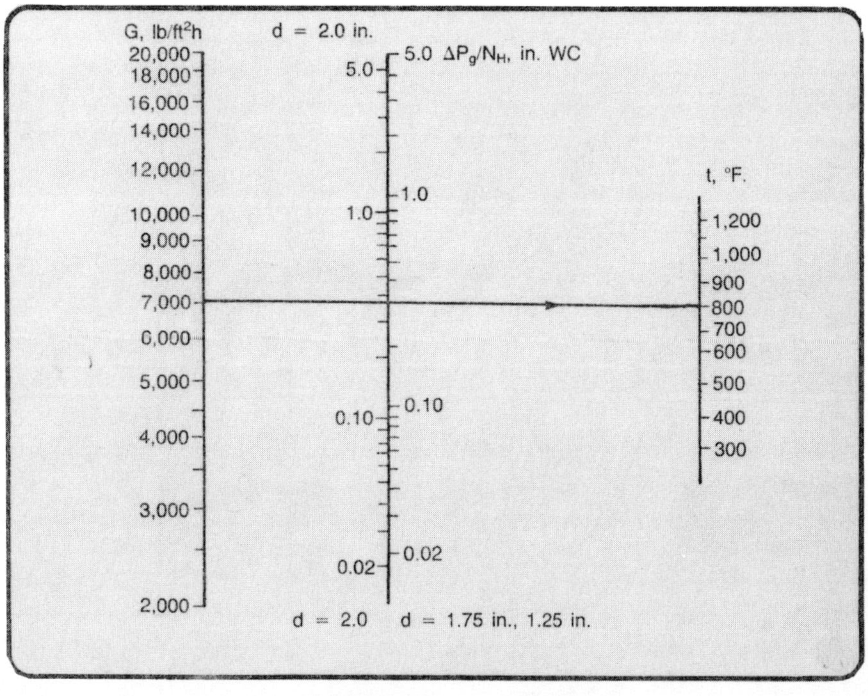


**Fig. 6. Gas-pressure drop in finned tubes**

Gas viscosity $= 0.075$ lb/ft h (see Fig. 2) and $MW = 30$. $\Delta P_g = 1.58 \times 10^{-8} \times 7{,}000^{1.684} \times 1.75^{0.611} \times 0.075^{0.316} \times 1{,}260 \times 6/(4.5^{0.927} \times 30) = 1.83$ in. WC.

### *Conclusion*

A few charts are presented to give gas-pressure drop over plain and finned tube bundles in cross flow. Wherever possible, graphs and data based on site-operational experience should be used. Generally, plant engineers and consultants may not have at hand such proprietary data so correlations and graphs presented here may be used for quick engineering estimates. They are easy to use and no reference to gas property data is necessary. Gas pressure drop information helps make useful recommendations on design.

### *References*

1. McAdams, W. J., "Heat Transmission," McGraw-Hill Book Co.
2. ESCOA Fin Tube Engineering Manual, Pryor, Okla., USA.
3. Ganapathy, V., "Charts simplify spiral finned-tube calculations," *Chemical Engineering,* April 25, 1977.
4. Extended Surface Tube Co. Catalog, U.K.

# 9. Chart relates pitot-tube readings to flow

THE ACCOMPANYING NOMOGRAPH PROVIDES A quick, simple method for estimating air velocity, density, and flow rate in air-conditioning and air-handling systems if the air temperature and a pitot-tube differential-pressure reading are known. Air density does not have to be calculated.

### *Example*

The pitot tube in a duct handling 200F air indicates a differential pressure of 0.4 in. w.g. The duct's cross-sectional area is 4 sq ft. What is the air's velocity, density, and flow rate?

### *Solution*

Align 200F on Scale T with 0.4 in. w.g. on Scale P and extend the line to the reference line. Where Scale V is crossed, read air velocity as 2900 fpm. Then, read air density as 0.06 lb/cu ft on Scale D.

Connect the marked point on the reference line with 4 sq ft on Scale A, and, where Scale Q is crossed, read air flow rate as 39,000 lb/hr.

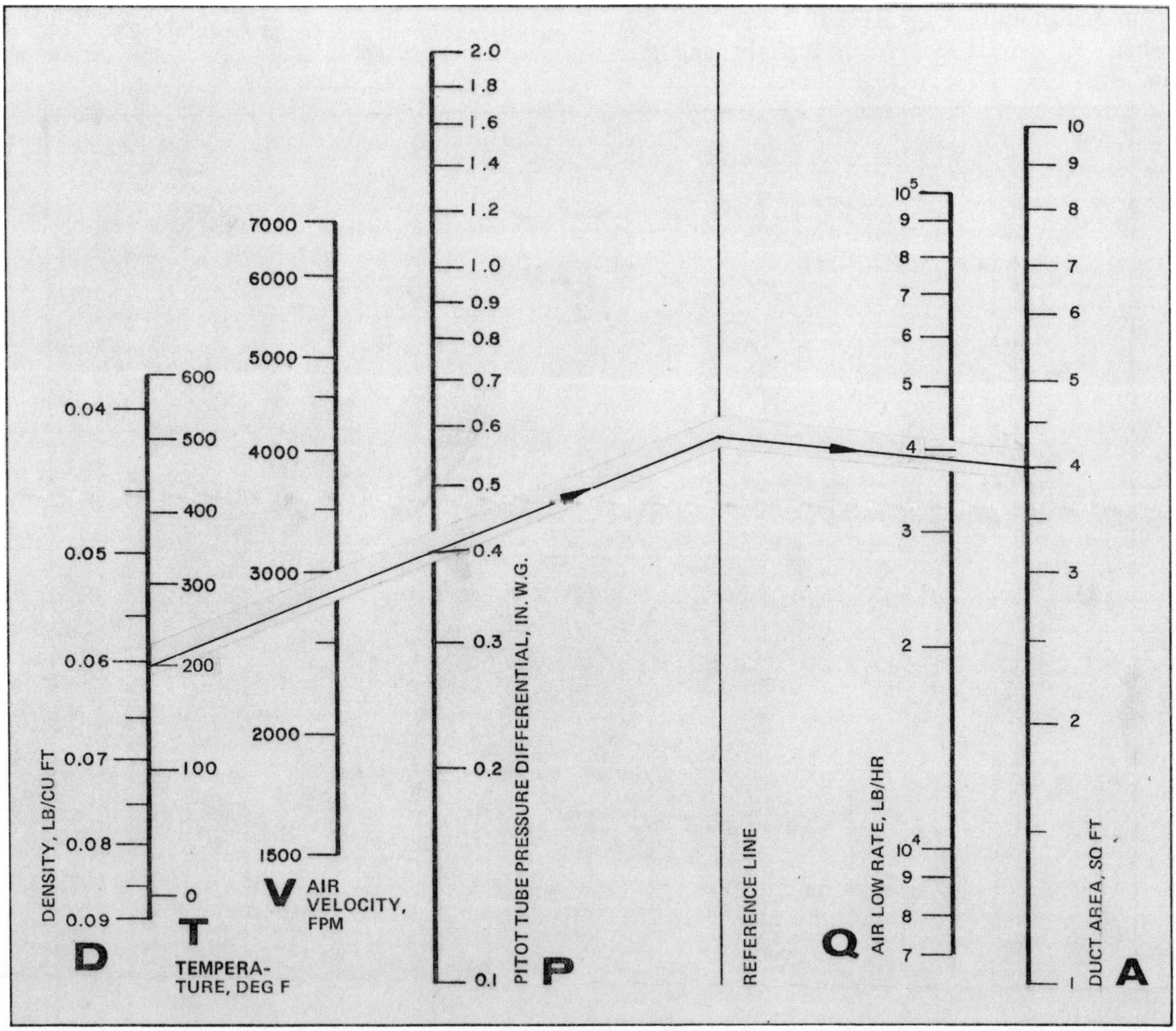

# 10. Chart relates gas-mass velocity with linear velocity

THE TERM "GAS-MASS VELOCITY" IS FREQUENTLY used in the design of heating surfaces of boilers, fired heaters, and the convective sections of waste-heat recovery equipment. "Linear velocity" in the section gives a good idea of fouling, erosion, and gas-pressure drop characteristics.

Though the conversion between the two terms is simple, the calculation or estimation of the flue-gas density makes it a bothersome affair.

To simplify the procedure, the following equations can be used:

In metric units: $V_g = 80 \times 10^{-8}\, GT$ (1)
where $V_g$ is in m/s; $G$ is in kg/m²hr; and $T$ is in °K.

In British units: $V_g = 710 \times 10^{-8}\, GT$ (2)
where $V_g$ is in ft/sec; $G$ is in lb/ft²hr; and $T$ is in °R.

In both equations, $G$ is the gas-mass velocity, $V_g$ is the linear velocity, and $T$ is the temperature.

These equations take into account the molecular weight and density of common flue gases and hence no computation of density is needed. For engineering estimates, the equations are reasonable. Also, the chart gives the conversion in British or metric units.

### *Example*

The chart shows that when gas-mass velocity $(G)$ is 40,000 kg/m²hr (8,000 lb/ft²hr), $T$ is 800°K (1,440°R), and $V_g$ is 26 m/sec (84 ft/sec).

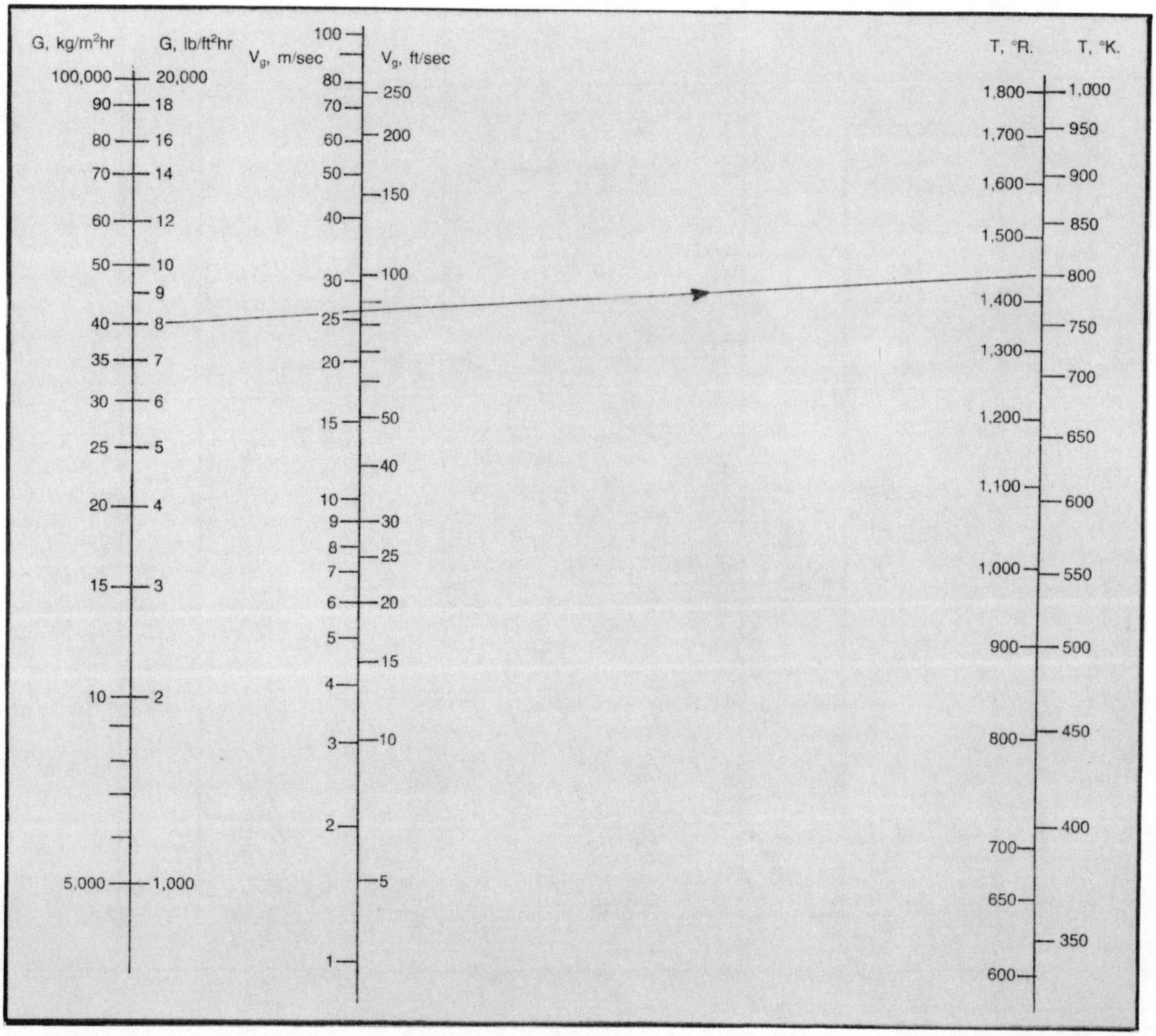

# 11. Charts estimate superficial-velocity boundaries in fluidized beds

THE BASIC DESIGN PARAMETER IN SIZING A fluidized bed is the gas superficial velocity. This value must be between the minimum fluidizing velocity and the terminal velocity above which a carryover of solids from the bed becomes possible.

Using the two nomograms presented here, a good estimation of these limiting velocities can be obtained if the density, mean particle size and sphericity of the solid, and the density and viscosity of the gas, are known.

The minimum fluidizing velocity, $U_m$, for Reynolds numbers of less than 20 to more than 1,000 is obtained from the following equations:[1]

$$Re_m = \left[33.7^2 + \frac{0.0408 d_p{}^3 \rho_g(\rho_s - \rho_g)g}{\mu^2}\right]^{1/2} - 33.7 \qquad (1)$$

where: $Re_m = 3{,}600\rho_g U_m d_p/\mu$

and: $U_m = Re_m \mu/3{,}600\rho_g d_p \qquad (2)$

The terminal velocity, $U_t$, is found by the following procedure. A dimensionless group, $C_d Re_t^2$, is calculated:

**Data on Sphericity**[1]

| Material | $\phi_s$ |
|---|---|
| Sand | 0.600, 0.861 |
| Iron catalyst | 0.578 |
| Bituminous coal | 0.625 |
| Celite cylinders | 0.861 |
| Broken solids | 0.63 |
| Sand | 0.534-0.628 |
| Silica | 0.554-0.628 |
| Pulverized coal | 0.696 |

Fig. 1

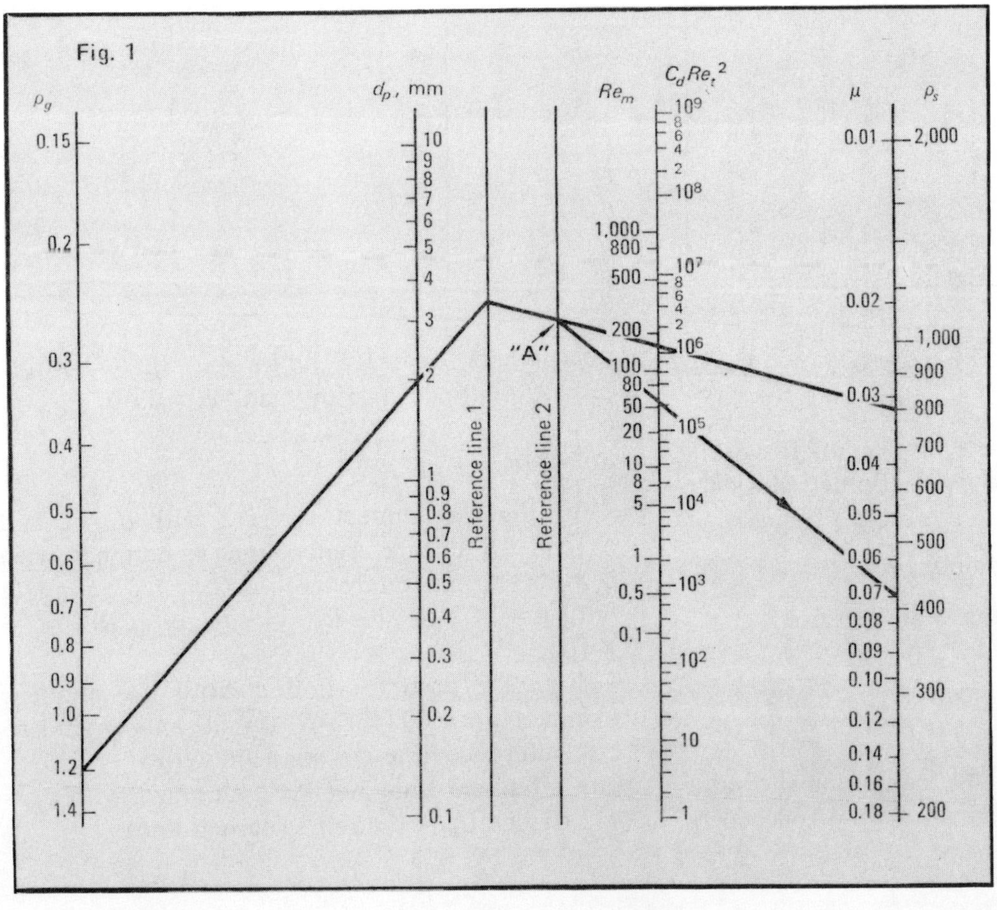


*Nomenclature*

| | |
|---|---|
| $d_p$ | Mean particle size, m |
| $\rho_g$ | Density of gas kg/m³ |
| $\rho_s$ | Density of solid kg/m³ |
| $\mu$ | Gas viscosity, kg/m-h |
| $Re_m$ | Reynolds number at minimum velocity |
| $Re_t$ | Reynolds number at terminal velocity |
| $U_m$ | Minimum superficial velocity, m/s |
| $U_t$ | Maximum superficial velocity, m/s |
| $C_d$ | Drag coefficient |
| $\phi_s$ | Sphericity of particles |
| $g$ | Acceleration of gravity, m/s² |

Fig. 2

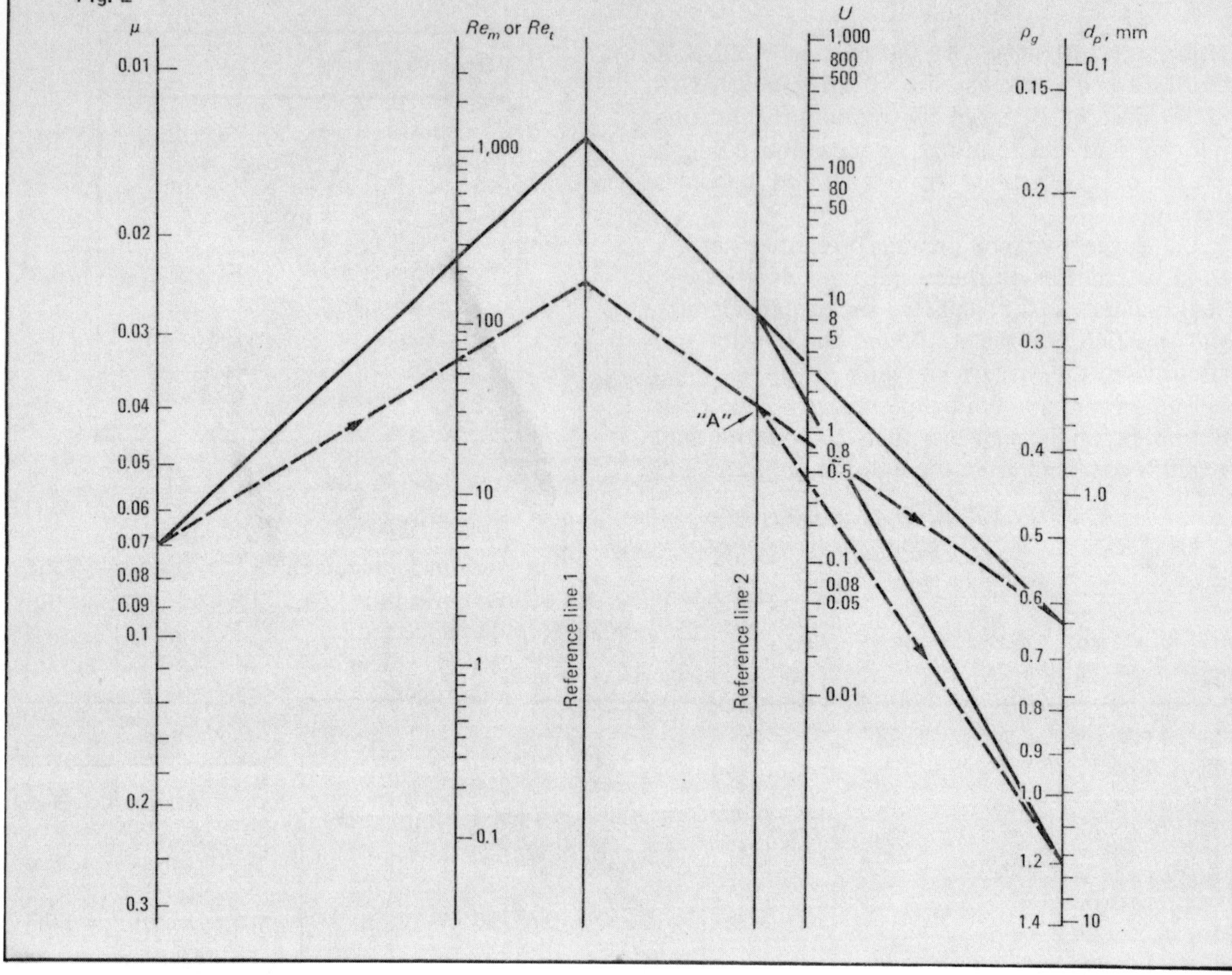


$$C_d Re_t^2 = \frac{4}{3} g\, d_p{}^3 \frac{\rho_g(\rho_s - \rho_g)}{\mu^2} \qquad (3)$$

Referring to Fig. 3, $Re_t$ is found, if $\phi_s$, the particle sphericity, is known. (Some values for $\phi_s$ are given in the Sphericity Table.) Hence:

$$U_t = Re_t \mu / 3{,}600 \rho_g d_p \qquad (4)$$

Note: The $C_d$-scales in Fig. 3 are not used.

### Example

Find the minimum and maximum allowable superficial velocity of gas in a fluidized bed of coal where: $\rho_g = 1.2$ kg/m³, $\mu = 0.07$ kg/m-h, $\rho_s = 800$ kg/m³, $d_p = 2$mm and $\phi_s = 0.60$.

### Solution

In Fig. 1, connect $\rho_g = 1.2$ with $d_p = 2$ and extend to cut reference line 1; connect with $\rho_s = 800$ to cut reference line 2 at "A"; connect with $\mu = 0.07$ to give $Re_m = 60$ and $C_d\ Re_t^2 = 2.5 \times 10^5$.

Fig. 2 is used for finding both $U_m$ and $U_t$. Connect $\mu = 0.07$ with $Re_m = 60$ and extend to cut reference line 1; connect with $d_p = 2.0$ to cut reference line 2 at "A". Connect with $\rho_g = 1.2$ to give $U_m = 0.45$ m/s (dashed line).

Fig. 3

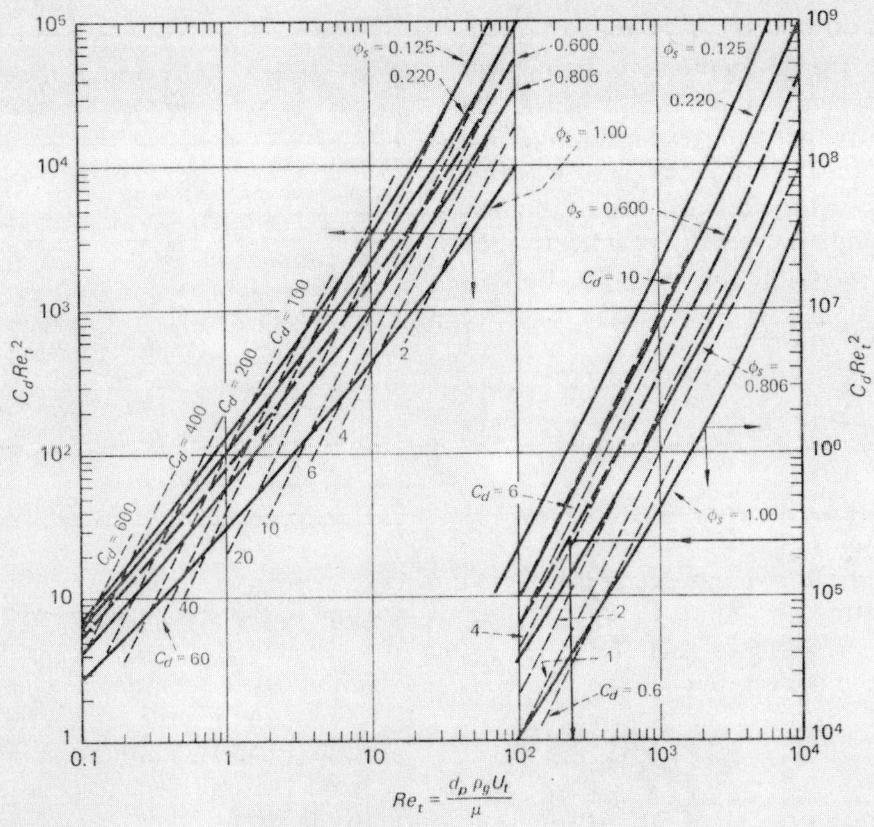


From Fig. 3, read $C_d\ Re_t^2 = 2.5 \times 10^5$ and $\phi_s = 0.60$ to find $Re_t = 250$. Following a similar procedure as with $Re_m$ from Fig. 2, get $U_t = 1.8$ m/s (continuous line).

### Reference

1. Kuni, I. D., and Levenspiel, O., *Fluidization Engineering,* John Wiley & Sons, 1969, pp. 65-77.

# 12. Nomograph estimates fuel oil flow and density at different temperatures

CONVERSION OF FUEL OIL FLOW FROM LB PER hr to gpm at different temperatures involves calculation of specific volumes, $v_t$, which depend on fuel oil gravity and expansion factors. The nomograph on the following page does this conversion without the need to look up the expansion factor and the specific volume at 60F. The nomograph solves the following equations:

$$W = 8Q/v_t \quad (1)$$

$$v_t = v_{60}\,[1 + E(t - 60)] \quad (2)$$

$$\rho = 1/7.48 v_t \quad (3)$$

where

$W$ = fuel oil flow, lb per hr
$Q$ = flow, gpm
$v$ = fuel oil specific volume, cu ft per lb
$t$ = temperature, F
$e$ = function of API (see Table 1)
$\rho$ = density, lb per gal.

### *Example*

Seven thousand lb per hr of 30° API fuel oil is heated from 60 to 150F. What is its specific volume, volume, and density given by Equation 3?

### *Solution*

Draw a vertical line from 150F on the $t$ scale up to 30° API. From this point, draw a horizontal line to the $\rho$ scale and read 7 lb per gal. as the density and 0.019 cu ft per lb as the specific volume (on the $v$ scale). Connect this point with 7000 on the $W$ scale and extend the line to the $Q$ scale. Read 16.5 gpm as the flow.

Note that the $W$ and $Q$ scales have the same multiplication factors, enabling the user to extend the range of the chart.

**Table 1. Coefficient of Thermal Expansion for Petroleum Oils**

| °API at 60F | $E$ coefficient (per F) |
|---|---|
| Below 14.9 | 0.00035 |
| 15.0-34.9 | 0.00040 |
| 35.0-50.9 | 0.00050 |
| 51.0-63.9 | 0.00060 |
| 64.0-78.9 | 0.00070 |
| 79.0-88.9 | 0.00080 |
| 89.0-93.9 | 0.00085 |
| 94.0-100.0 | 0.00090 |

### *Bibliography*

*North American Combustion Handbook,* 2nd edition, p. 23.

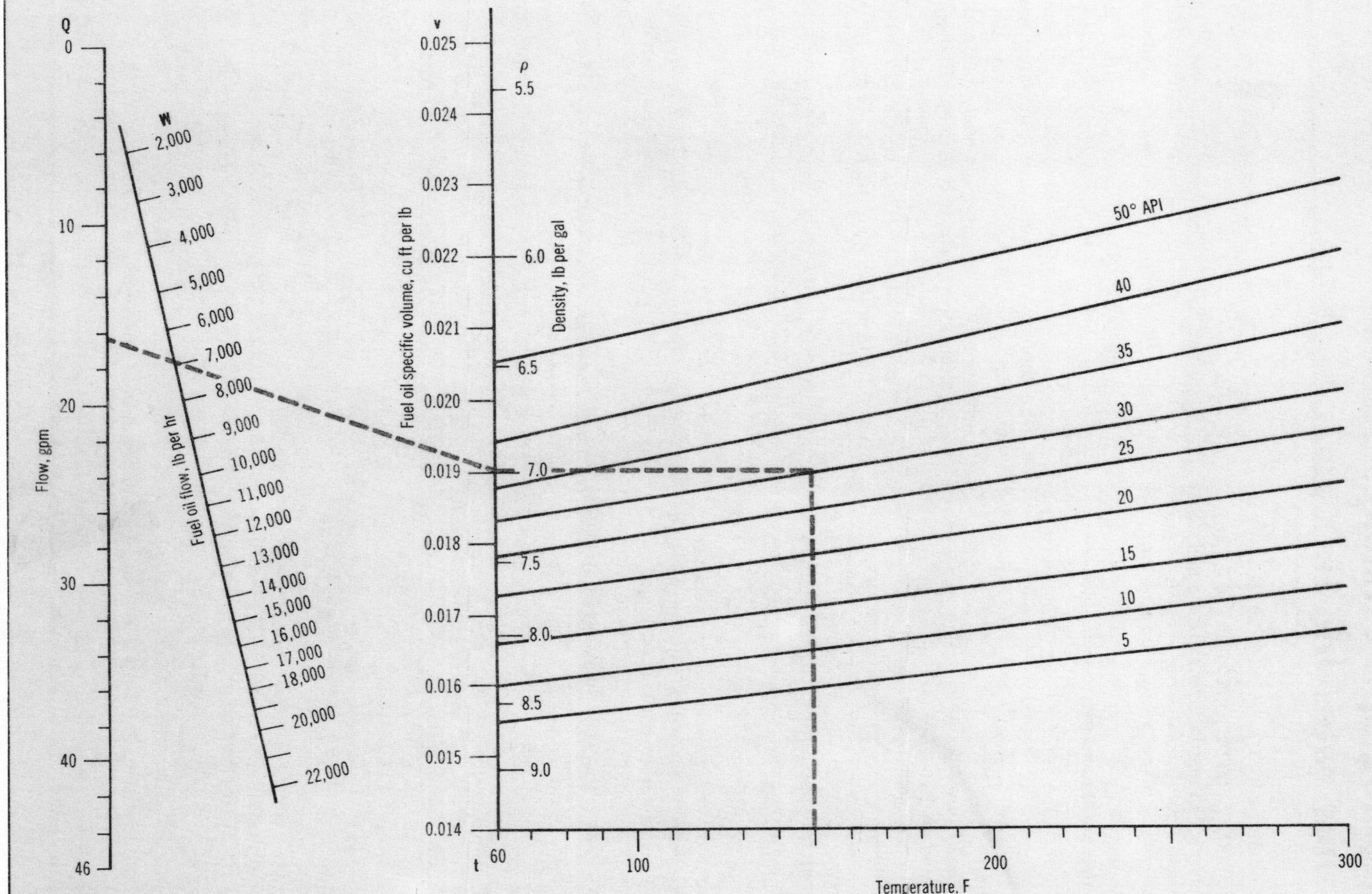


Nomograph estimates fuel oil flow and density at different temperatures. Broken lines indicate solution to example in text.

# 13. Nomograph estimates critical flow rates for air

CRITICAL FLOW CONDITIONS FOR AIR ARE found in several industrial applications such as flow through soot blower nozzles, spray guns, safety valves, and leakage through holes from pressure vessels.

The expression that relates the variables is:[1]

$$W = 356AP(M/T)^{1/2} \qquad (1)$$

where

$W$ = flow, lb per hr
$A$ = area of opening, sq in.
$M$ = molecular weight of air
$T$ = absolute temperature, R
$P$ = relieving or discharge pressure, psia

The above equation can be simplified and rewritten after substituting $M = 29$ and $T = 530$ R:

$$W = 83.3AP \qquad (2)$$

The effect of variations in air temperature is marginal because $T$ is part of the square root quantity. For example, between 40 and 80F, the variation would be $[(460+40)/(460+80)]^{1/2} = 0.96$ or 4 percent. Hence, we may neglect the ambient air temperature variations.

### *Example*

What is the leakage air flow from a pressure vessel at 40 psia if the hole is ¼ in. in diameter?

### *Solution*

The nomograph (on the following page) solves Equation 2. It gives $W$ in lb per hr as well as the flow in scfm. Also the area can be in sq in. or, alternatively, the hole diameter in in. may be used.

Connect 40 on the $P$ scale with ¼ on the $d$ scale. Extend the line to the $W$ scale and read $W = 170$ lb per hr and $Q = 35$ scfm.

Scales $A$, $W$, and $Q$ have the same multiplication factor, thus the range of the chart can be extended.

### *Reference*


*ASME Boiler and Pressure Vessel Code*, Section VIII, Div. 1, Para. UG 131.

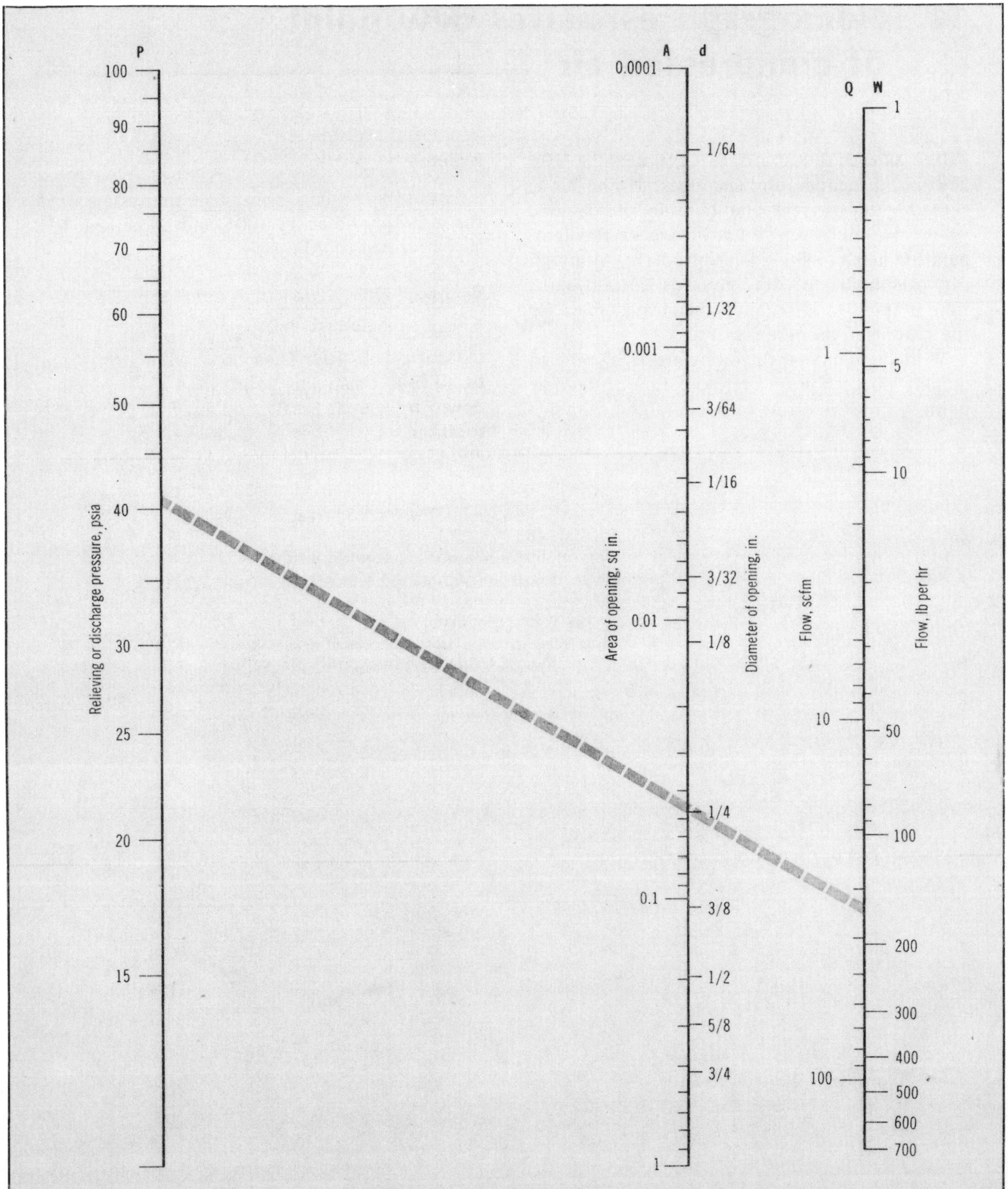


Nomograph estimates critical flow rates for air. Broken line indicates solution to example in text.

# 14. Nomograph estimates dew point of compressed air

WHEN ATMOSPHERIC AIR IS COMPRESSED, THE dew point temperature increases. If the air is cooled below this temperature, as in intercoolers, water can condense. A knowledge of this temperature helps to size coolers properly. Although the calculation of dew point is not difficult, time is spent in looking up the steam tables for the saturation properties of water.

With the chart on the following page, one can quickly obtain a close approximation of the dew point without referring to steam tables.

### *Example*

Estimate the dew point of compressed air if the atmospheric air at 100F and 90 percent RH is compressed to 80 psia.

### *Solution*

Draw a vertical line from 100 on the $T_a$ scale to 90 percent RH. Then draw a horizontal line to 80 psia. From this point, draw a vertical line down to the $T_d$ scale and read 159F as the solution.

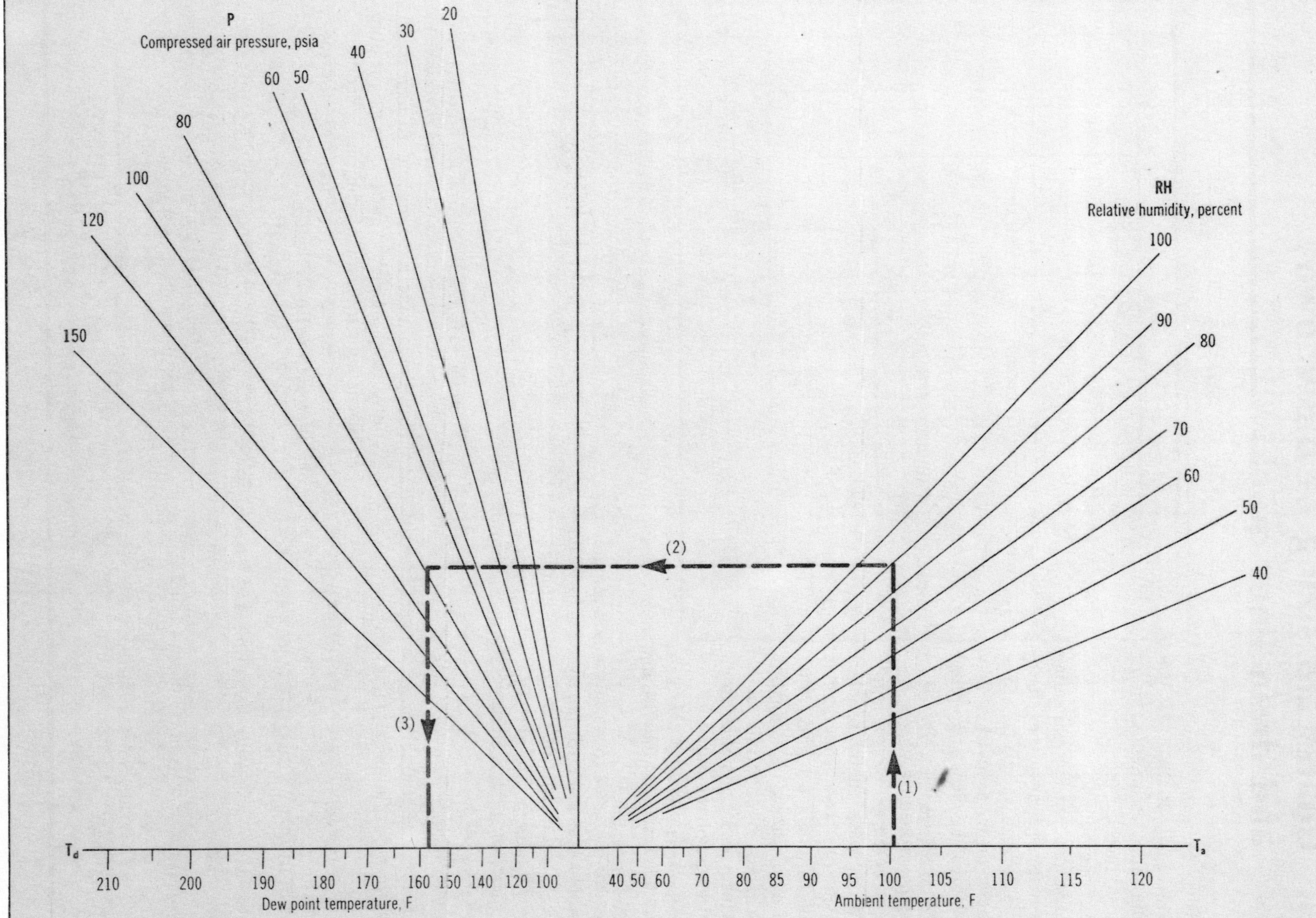


Nomograph estimates dew point of compressed air. Broken lines indicate solution to example in text.

# 15. Charts aid sizing of blowdown and drain lines quickly

SIZING OF BLOWDOWN OR drain lines assumes a lot of importance in boiler or process plant operations.

The problem of estimating the discharge rates from a boiler drum or vessel to atmosphere or to a vessel at low pressures involves two-phase flow calculations and is a lengthy procedure.[1,2]

Presented below is a simplified approach to the problem which can save considerable time for engineers who are involved in sizing or estimating discharge rates from boiler drums, vessels, or similar applications involving water.

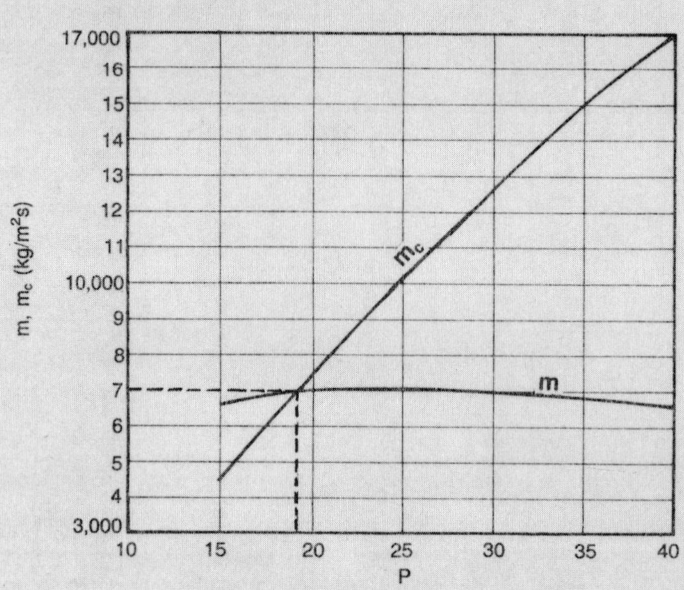

**Fig. 1. Calculation results**

Several advantages are claimed for these charts, including:

- No reference to steam tables is required.
- No trial and error procedure is involved.
- Effect of friction can be easily studied.
- Obtaining pipe size to discharge a desired rate of fluid, the reverse problem, is simple.

### *Theory*

The basic Bernoulli's equation may be written as follows for flow in a piping system:

$$10^4\, \nu dP + \frac{\mathrm{v}^2}{2g} dk + \frac{\mathrm{v}}{g} d\mathrm{v} + dH = 0 \qquad (1)$$

Substituting mass flow rate $m = \mathrm{v}/\nu$:

$$\frac{m^2}{2g}\left(dk + 2\frac{d\nu}{\nu}\right) = -10^4 \frac{dp}{\nu} - \frac{dH}{\nu^2} \qquad (2)$$

Integrating between conditions 1 and 2:

$$\frac{m^2}{2g}\left[k + 2 ln \frac{\nu_2}{\nu_1}\right] = -10^4 \int_1^2 \frac{dp}{\nu} - \int_1^2 \frac{dH}{\nu^2} \qquad (3)$$

or:

$$m = \left[\frac{2g}{\left(k + 2\, ln \frac{\nu_2}{\nu_1}\right)} \times \left[-10^4 \int_1^2 \frac{dp}{\nu} - \int_1^2 \frac{dH}{\nu^2}\right]\right]^{1/2} \qquad (4)$$

Where $K = (fl/d)$, the equivalent pipe resistance.

When the pressure of the vessel to which the blow-down pipe is connected is decreased, the flow rate increases until critical pressure is reached at the end of the pipe. Reducing the vessel pressure below critical pressure does not increase the flow rate.

If the vessel pressure is less than the critical pressure, critical flow conditions are reached and sonic flow results.

From thermodynamics, the sonic velocity may be shown to be as follows:

$$V_c = \sqrt{-\nu^2 g \left(\frac{dP}{d\nu}\right)_s \times 10^4} \qquad (5)$$

### Nomenclature

- $d$ inner diameter of pipe, m
- $f$ friction factor
- $g$ acceleration due to gravity, $m/s^2$
- $h$ enthalpy, Kcal/kg
- $H$ elevation, m
- $I$ equivalent length of pipe, m
- $k$ $(fl/d)$, the coefficient of resistance
- $m$ mass flow rate of steam-water mixture
- $p$ pressure of fluid, atm
- $S$ entropy, Kcal/kg°C
- $v$ specific volume, $m^3/kg$
- v velocity, m/s
- $\rho$ density, $kg/m^3$
- $x$ dryness fraction

### Subscript

- $l$ liquid
- $v$ vapor
- $c$ critical
- $s$ saturated liquid
- $m$ mean

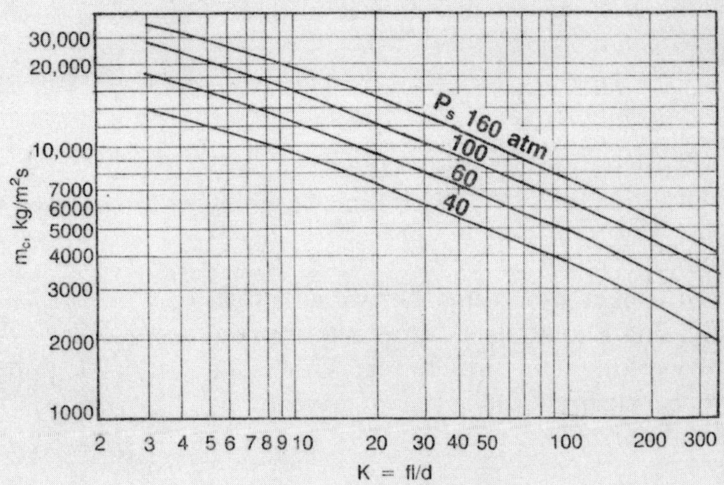


**Fig. 2. Solving for *m***

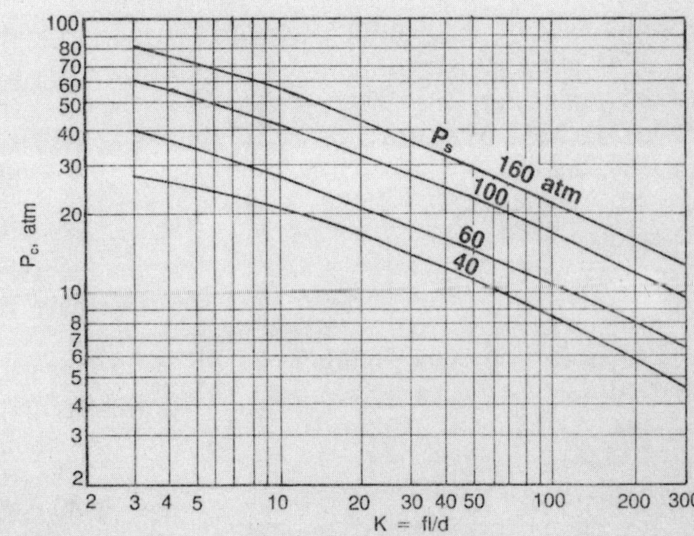


**Fig. 3. Solving for *P***

$$m_c = 100\sqrt{-g\left(\frac{dP}{dv}\right)_s} \qquad (6)$$

The term $(dp/dv)_s$ refers to the change in pressure to volume ratio at critical flow conditions at constant entropy.

Hence, in order to estimate $m_c$, Equations 4 and 6 have to be solved. This is an iterative procedure and the steps are outlined below. For the sake of simplicity the term involving the height differences will be neglected. For high pressure systems the error in neglecting this term is marginal, on the order of 5%.

The problem is, then, given $K$ and $P_s$, to estimate $P_c$ and $m$. This is a trial and error procedure and the steps are outlined below, followed by an example. Above, two charts are presented which may be used for quick sizing purposes.

1. Assume $P_c$.
2. Calculate $(dp/dv)_s$ at $P_c$ for constant entropy conditions. The volume change corresponding to 2 to 3% of $P_c$ may be calculated and then $(dp/dv)_s$ may be obtained.
3. Calculate $m_c$ using Equation 6.
4. Equation 4 has to be solved for $m_c$.

The term $[-10^{-4} \int^2 dp/v]$ is computed as follows using Simpson's rule:

$$-10^4 \int_1^2 \frac{dP}{v} = -10^4 \int_1^2 \rho dP$$

$$= \frac{(Ps-Pc)}{6} [\rho_s + 4\rho_m + \rho_c] \qquad (7)$$

$\rho_m$ is the density at mean pressure of $(P_s+P_c)/2$.

The densities are computed at isenthalpic conditions. The term $2\ ln\ (v_2/v_1) = 2\ ln\ (\rho_s/\rho_c)$ is then found.

Then, $m$ is computed using Equation 4. If $m$ computer using Equations 4 and 6 tally, assume $P_c$ and the resultant $m_c$ are correct. Otherwise $P_c$ has to be changed and all steps have to be repeated until $m$ and $m_c$ agree.

### *Example*

A boiler drum blowdown line is connected to a tank set at 8 atm. Drum pressure is 100 atm and the resistance $K$ of the blowdown line is 80. Estimate the critical mass flow rate $m_c$ and the critical pressure $P_c$.

The procedure will be detailed for an assumed pressure $P_c$ of 40 atm.

For steam table $P_s$=100 atm, $s$=0.7983 (Kcal/kg°C); $h_l$=334 Kcal/kg, $v_l$=0.001445 m$^3$/kg or $\rho$=692 kg/m$^3$.

Let $\rho_c$=40 atm, then $h_l$=258.2 Kcal/kg, $h_v$=669 Kcal/kg, $S_l$=0.6649, $S_v$=1.4513, $v_l$= 0.001249, $v_v$=0.05078,

Hence:

$$x = \frac{S-S_l}{S_v-S_l} = \frac{0.7983-0.6649}{1.4513-0.6649}$$

$$= 0.1696$$

$$v = v_l + x(v_v - v_l)$$

$$= 0.001249+0.1696 \times (0.05078- 0.001249) = 0.009651 \text{ m}^3/\text{kg}$$

Again, compute v at 41 atm (2.5% more than $P_c$). Using the similar steps as described above, v=0.0093 m$^3$/kg.

Hence:

$$m_c = 100\sqrt{-g\left(\frac{dP}{dv}\right)_s}$$

$$= 100\sqrt{\frac{9.8 \times 1}{(0.00965-0.0093)}}$$

$$= 16733 \text{ kg/m}^2\text{s}$$

Compute the densities as shown below:

$$\rho_s = \frac{1}{0.001445} = 692 \text{ kg/m}^3$$

The dryness fraction at 40 atm at isenthalpic condition is:

$$x = \frac{334-258.2}{669-258.2} = 0.1845$$

$$v_c = 0.001249+0.1845\ (0.05078-0.001249)$$
$$= 0.010387 \text{ m}^3/\text{kg}$$
$$\rho_c = 96.3 \text{ kg/m}^3$$

Similarly, at $P_m$ =(100+40)/2=70 atm:

$$v_m = 0.03785 \text{ m}^3/\text{kg or}$$
$$\rho_m = 264 \text{ kg/m}^3$$

$$-10^4 \int_1^2 \frac{dP}{v} = \frac{(100-40)}{6} (692+4 \times 264+96.3)$$

$$= 184 \times 10^6\ \ 2ln(v_2/v_1) = 2ln(\rho_s/\rho_c) = 4.6$$

Substituting the various quantities in Equation 4:

$$m = \sqrt{\frac{2 \times 9.8}{(80+4.6)} \times 184 \times 10^6}$$
$$= 6530 \text{ kg/m}^2\text{s}$$

The two values $m$ and $m_c$ do not agree. Hence we have to repeat the calculations for another $P_c$.

This has been done for $P_c$=30 and 15 and the results are presented in Fig. 1.

At about 19 atm, the two curves intersect and the mass flow rate is about 7,000 kg/m$^2$ s.

However one may do the calculations at this pressure and check.

### *Use of Charts*

As seen above, the procedure involved is lengthy, tedious, and trial and error is involved. Also, reference to steam tables makes it a cumbersome procedure. Hence from various $K$ values and initial pressure $P_s$, a calculator was used to solve for $P_c$ and $m$ and the results are presented in Figs. 2 and 3.

These charts are very simple to use and results are obtained at a glance for various $K$ values. These two charts are extremely handy when it comes to estimating $m_c$, given $P_s$ and $K$; or conversely, $K$, given $P_s$ and a desired $m_c$.

### *References*

1. F. J. Moody, "Maximum two-phase vessel blow down from pipes," Trans ASME, *J of Heat Transfer,* Aug. 1966, p. 285.
2. F. J. Moody, "Maximum flow rate of a single component two-phase mixture," Trans ASME, *J of Heat Transfer,* Feb. 1965, p. 53.

# 16. Chart estimates leakage across dampers and cost

DAMPERS ARE USED IN GAS AND AIR DUCTING IN POWER PLANTS, process industries, and air conditioning systems to regulate or bypass flow.

One important application is in combined cycle plants where the turbine exhaust gases are bypassed to the stack when the boiler is not used (Fig. 1). When the boiler is in operation it is assumed that all gas from the gas turbine goes to the boiler.

However, this is an ideal situation, as any damper usually involves some leakage.

The energy loss incurred due to the leakage across the damper may be significant and plant engineers would do well to be judicious in

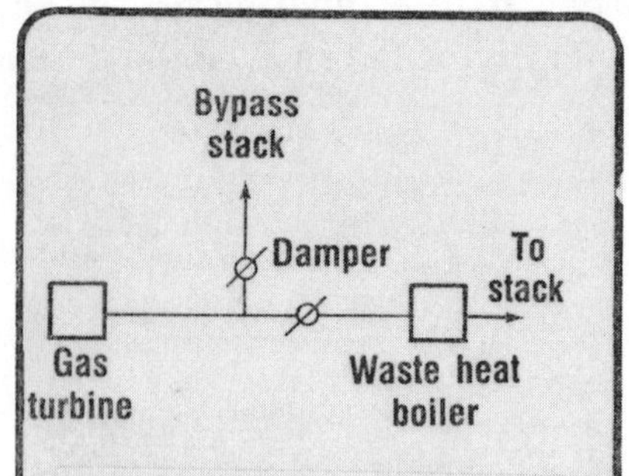


**Fig. 1. Damper arrangement**

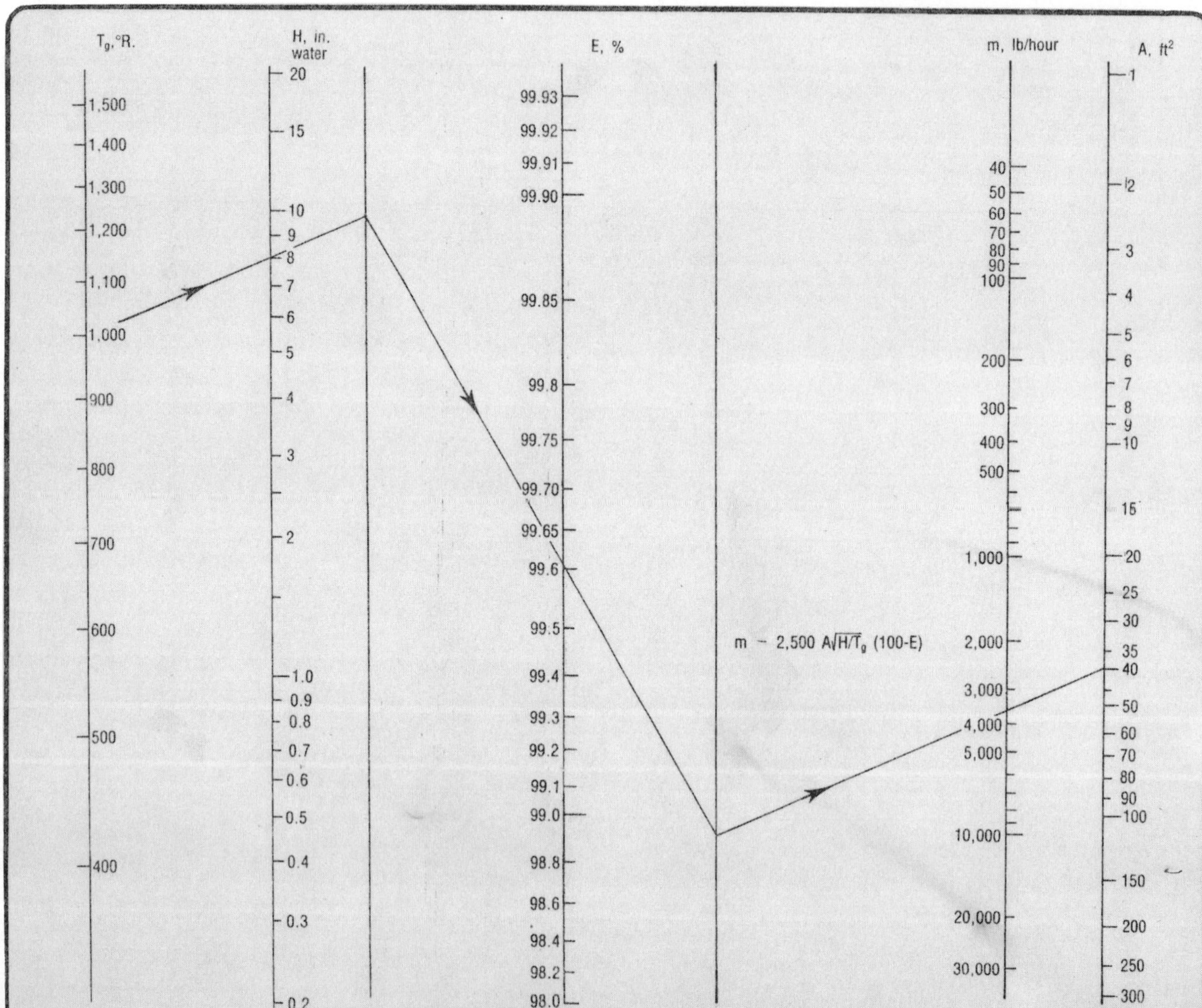


**Fig. 2 Example solution**

### *Nomenclature*

| | |
|---|---|
| $A$ | Area of damper, cross section, $ft^2$ |
| $C_d$ | Coefficient of discharge |
| $C_a$ | Annual cost of energy loss due to leakage across dampers, \$ |
| $C_p$ | Specific heat of gas, Btu/lb°F |
| $e$ | Escalation |
| $E$ | Efficiency of sealing of dampers, % |
| $F$ | Capitalization factor |
| $H$ | Differential pressure across damper, in., water column |
| $i$ | Interest rate |
| $g$ | Acceleration due to gravity, $ft/s^2$ |
| $m$ | Leakage rate, lb/hr |
| $N$ | Life of the plant, years |
| $P_g$, $P_w$ | Densities of gas and water, lb/cu ft |
| $Q$ | Energy loss, Btu/hr |
| $V$ | Velocity of leakage, ft/s |
| $t_g$, $t_o$ | Temperatures of gas and ambient condition, °F |
| $T_g$ | Gas temperature, °R |
| $T$ | Total capitalized cost |

selecting dampers and understand the impact of leakage. This is indirectly expressed as the efficiency of sealing.

In what follows, an expression is derived to quantify the leakage and compute the associated energy loss and the cost thereof.

### *Leakage, Energy Loss*

Let $E$ be the efficiency of gas tightness or sealing efficiency of the damper. This depends on the size of the duct. For small ducts, this can be very high (over 99.9%), but for large ducts it could be lower (99%). Damper manufacturers provide or guarantee such figures.

If $H$ is the differential across the damper in inches of water column, the velocity of gas or air leaking across is:

$$V = C_d \sqrt{\frac{2gH}{2} \frac{P_w}{P_g}} \qquad (1)$$

Then, $m$, the quantity of gas leaking is:

$$m = 3600\, P_g\, A\, V \left(\frac{100\text{-}E}{100}\right) \qquad (2)$$

Using a value of 0.6 for $C_d$ and a molecular weight of 29.5 for the flue gas, we may rewrite Equation 2 after substituting for $V$ and $P_g$, as follows:

$$m = 2500\, A \sqrt{\frac{H}{T_g}}\, (100\text{-}E) \qquad (3)$$

The energy loss is then:

$$Q = m\, C_p (t_g - t_o) \qquad (4)$$

Fig. 2 solves for $m$. Once $m$ is computed, $Q$ and annual energy loss may be obtained.

### *Example*

An isolation damper is used to shut off hot gases entering the bypass stack of a boiler in a combined cycle plant as shown in Fig. 1.

Area of the duct, $A = 40\ ft^2$
Gas temperature, $T_g = 540°F = 1000°R$
Pressure differential, $H = 8$ in. WC
Ambient temperature, $t_o = 80°F$
Efficiency of sealing of damper, $E = 99.6\%$

Estimate the leakage across the damper to the atmosphere; energy loss in Btu/hr; and the annual cost of leakage, if plant works for 6000 hrs and energy costs \$4/MMBtu.

### *Solution*

In Fig. 2, connect $T_g = 1000°R$ with $H = 8$ and extend to cut line 1. Connect with $E = 99.6\%$ and extend to cut line 2. Connect with $A = 40$ to cut $m$ scale. Read off 3,600 lb/hr.

The energy loss, if specific heat is taken as 0.25 Btu/lb°F, is:

$$Q = 3600 \times 0.25 \times (540 - 80) = 0.41 \times 10^6 \text{ Btu/hr}$$

$$C_a = \text{Annual energy loss} = 0.41 \times 10^6 \times 6000 \times {}^4/_{10^6} = \$9{,}480$$

### *Evaluation of Alternatives*

Since even a 0.1% change in $E$ means a lot by way of savings in energy costs, efforts should be made to evaluate various bids in terms of life cycle costs. Assume, for example, another damper manufacturer guarantees 99.9% of sealing efficiency but quotes \$95,000 against \$45,000 for the case of 99.6% efficiency damper. Which do we choose?

Life cycle costing may be applied. If $e$ is the escalation and $i$, the interest rate, the capitalized cost $T$ of an alternative is given by:

$$T = \text{Fixed cost} + C_a \left(\frac{1+e}{1+i}\right) \left[\frac{1-\left(\frac{1+e}{1+i}\right)^N}{1-\left(\frac{1+e}{1+i}\right)}\right] \tag{5}$$

$$= \text{Fixed cost} + C_a \times F$$

where $N$ is the life of the plant.

Assuming $e=6\%$, $i=9\%$, $N=20$ years, the total capitalized cost for both bids is shown in Table 1.

**Table 1. Capitalized Costs**

| | No. 1 | No. 2 |
|---|---|---|
| Fixed cost, \$ | 45,000 | 95,000 |
| Efficiency, $E$, % | 99.6 | 99.9 |
| $C_a$, \$ | 9840 | 2460 |
| $F$ | 15.1 | 15.1 |
| $C_a \times F$, \$ | 148,931 | 37,233 |
| $T$, \$ | 193,931 | 132,233 |

It is obvious that efficiency guaranteed makes a lot of difference; a higher efficiency in spite of higher investment seems to be economical in the long run.

Leakage across dampers is an important factor which should be considered while evaluating a bid. Equations to determine the leakage rate, energy loss, and the capitalized cost of the damper operation using life cycle costing technique give a good idea of the economical alternative.

As fuel costs increase, the trend will be to go to costlier dampers with better sealing efficiency.

# Appendix: Conversion factors

| To obtain | Multiply | By |
|---|---|---|
| atmospheres | psi | 0.068 |
| Btu | Kcal | 3.9685 |
| Btu | Kwh | 3413 |
| Btu/ft$^2$ hr | Kcal/m$^2$ hr | 0.369 |
| Btu/lb | Kcal/kg | 1.8 |
| Btu/lb °F | Kcal/kg °C | 1.0 |
| Btu/ft$^2$ hr °F | Kcal/m$^2$ hr °C | 0.204 |
| Btu/ft hr °F | Kcal/m hr °C | 0.671 |
| Btu/cu ft | Kcal/cu m | 8.92 |
| centimeter | in. | 2.54 |
| cm of Hg at 0°C | in. $H_2O$ at 4°C | 0.1868 |
| cm of Hg at 0°C | psi | 5.1715 |
| centistokes | centipoises | 1/density |
| cu ft | cu meters | 35.31 |
| cu meters | gal. (USA, liq.) | 0.003785 |
| dyne-cm | ft-lb | 13,558,000 |
| ergs | Joules | 10,000,000 |
| feet | meters | 3.28 |

| To obtain | Multiply | By |
|---|---|---|
| gal. | cu ft | 7.48 |
| gal. | cu meter | 264.1 |
| grains | lbs | 7000 |
| Joules | calories | 4.186 |
| kg | lbs | 0.4536 |
| kg/cu meter | lb/cu ft | 16.02 |
| Kcal/m hr °C | Btu/ft hr °F | 1.49 |
| kg/m h | lb/ft hr | 1.49 |
| kg/sq cm | atmospheres | 1.033 |
| kg/sq cm | psi | 0.0703 |
| kg/sq ft | lb/sq ft | 4.88 |
| Kw | hp | 0.746 |
| Kw | Kcal/hr | 0.001163 |
| pounds | tons, metric | 2204.6 |
| pounds | tons, long | 2240 |
| pounds/cu ft | kg/cu meter | 0.0624 |
| lb/ft hr | centipoises | 2.42 |

# Index